Extragalactic Novae

A historical perspective

IOP Astronomy

AAS Editor in Chief

Ethan Vishniac, John Hopkins University, Maryland, US

About the program:

AAS-IOP Astronomy ebooks is the official book program of the American Astronomical Society (AAS), and aims to share in depth the most fascinating areas of astronomy, astrophysics, solar physics and planetary science. The program includes publications in the following topics:

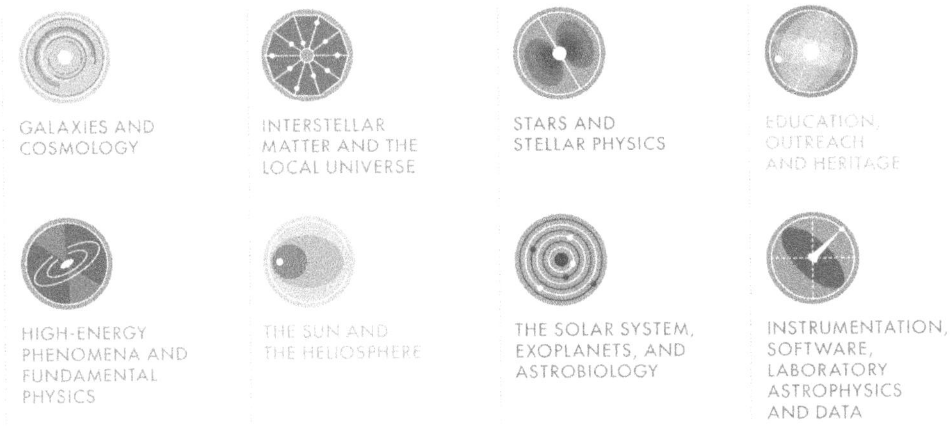

Books in the program range in level from short introductory texts on fast-moving areas, graduate and upper-level undergraduate textbooks, research monographs and practical handbooks.

For a complete list of published and forthcoming titles, please visit iopscience.org/books/aas.

About the American Astronomical Society

The American Astronomical Society (aas.org), established 1899, is the major organization of professional astronomers in North America. The membership (~7,000) also includes physicists, mathematicians, geologists, engineers and others whose research interests lie within the broad spectrum of subjects now comprising the contemporary astronomical sciences. The mission of the Society is to enhance and share humanity's scientific understanding of the universe.

Extragalactic Novae

A historical perspective

Allen W Shafter
San Diego State University, San Diego, CA, USA

IOP Publishing, Bristol, UK

ISBN 978-0-7503-1296-7 (ebook)
ISBN 978-0-7503-1297-4 (print)
ISBN 978-0-7503-1298-1 (mobi)

DOI 10.1088/2514-3433/ab2c63

Version: 20190901

AAS–IOP Astronomy
ISSN 2514-3433 (online)
ISSN 2515-141X (print)

British Library Cataloguing-in-Publication Data: A catalogue record for this book is available from the British Library.

Published by IOP Publishing, wholly owned by The Institute of Physics, London

IOP Publishing, Temple Circus, Temple Way, Bristol, BS1 6HG, UK

US Office: IOP Publishing, Inc., 190 North Independence Mall West, Suite 601, Philadelphia, PA 19106, USA

This book is dedicated to the work and the memory of my colleague and long-time friend, Dr. John Kendall Cannizzo (1957–2018).

"I like the stars. It's the illusion of permanence, I think. I mean, they're always flaring up and caving in and going out. But from here, I can pretend … I can pretend that things last. I can pretend that lives last longer than moments. Gods come, and gods go. Mortals flicker and flash and fade. Worlds don't last; and stars and galaxies are transient, fleeting things that twinkle like fireflies and vanish into cold and dust. But I can pretend …"

—***The Sandman: Brief Lives*** **© 1992 DC Comics.**
Written by Neil Gaiman.

Contents

Part II The Modern Era

Preface

Approximately two to three times per decade somewhere in the night sky a star suddenly reaches a brightness easily accessible to the unaided eye, before fading inexorably back into obscurity. These "guest stars," which have been observed since ancient times, eventually became known by the Latin name for "new stars," *Novae Stella*, or simply *Novae*. In contrast to the constancy of the multitudes of fixed stars, novae stand out as among the most spectacular events seen in the Heavens.

A major goal of this book is to give the reader an introduction to, and a broad historical perspective on, the role that the novae have played in the development of modern astronomy. The book begins with a brief overview of early nova observations, the initial confusion between novae and supernovae, and ultimately the role played by novae in the birth of extragalactic astronomy and cosmology. The discussion then turns to a more or less chronological account of extragalactic nova studies that took place throughout the remainder of the twentieth century and up through the early years of the new millennium.

Although quite enigmatic initially, novae were eventually understood to be powered by a thermonuclear runaway in the material accreted onto the surface of a white dwarf star from a close binary companion. At the peak of their eruptions novae become among the most powerful of celestial sources, with luminosities up to a million times that of the Sun.

It was recognized early on that the peak luminosities of novae were weakly correlated with their rate of decline from maximum light. Early attempts to calibrate this relationship, known as the "maximum-magnitude, rate-of-decline" (MMRD) relation were made in the hope that novae would provide a viable extragalactic distance indicator. Although that early promise has fallen short of expectations, their high luminosities and characteristic light curves make novae visible in galaxies up to 100 million light years distant. Thus, novae have the potential to be powerful probes of close binary star formation and evolution across a wide range of galaxy morphologies. The book concludes by chronicling the latest extragalactic nova studies: what we have learned, and what future studies may yet reveal, concerning extragalactic nova populations.

Acknowledgments

When I set out to write this book, I had little idea of the time and effort that would be involved in seeing it through to the end. The journey was made less arduous thanks to the help and encouragement of several people.

I'd like to begin by thanking my old friend and colleague Professor C. Theodore Daub for his many careful readings of the manuscript as it was being developed, and for his numerous suggestions that led to a real improvement in the presentation. Professors Summner Starrfield and Fred Talbert also provided valuable comments and suggestions for which I am grateful.

My colleagues Drs. Martin Henze and Matt Darnley deserve a special acknowledgement for their many contributions to the subject of extragalactic novae over the years, and for their detailed and insightful comments on the manuscript. I have learned a lot about the study of extragalactic novae from my close collaborations with both of them.

Finally, the book could not have been written without the love and support of my wife, Lani, and our three children, for which I'm most appreciative.

About the Author

Allen W Shafter

Allen W Shafter is a native of southern California who became interested in Astronomy at an early age after listening to the late Professor George Abell on local radio. Shafter did his undergraduate work in Physics at the University of California at Irvine, and subsequently received his doctorate in Astronomy from UCLA in 1983 working on the Cataclysmic Variable stars. His interest in the study in extragalactic novae, which was to consume most of his professional career, was kindled at UCLA through the early Hα imaging of M31 novae by Professor Holland Ford and his then graduate student Robin Ciardullo. Shafter has now been studying extragalactic novae for more than 35 years.

After leaving UCLA, Shafter spent six years at the University of Texas at Austin, initially as the recipient of a McDonald Observatory postdoctoral fellowship, and later as a Research Scientist working primarily with Professor Edward L. Robinson on various aspects of cataclysmic variable star research. He then returned to southern California in 1989, where he joined the Astronomy faculty at San Diego State University. Shafter served as Chairman of the Department from 2005 until 2017, before assuming his current position as Emeritus Professor of Astronomy.

During his career, Shafter has published more than 100 refereed papers, mostly on Cataclysmic Variable stars generally, and novae in particular. In addition to his faculty responsibilities, Shafter currently serves as a Scientific Editor for the Stars and Stellar Physics corridor of the AAS journals.

Part I

Foundations

IOP Astronomy

Extragalactic Novae
A historical perspective
Allen W Shafter

Chapter 1

The Early History of the Transient Sky

The wonder of the nighttime sky has been a source of awe and mystery since the dawn of mankind. According to the *Yale Bright Star Catalog*, there are almost 10,000 stars visible to the unaided eye (brighter than magnitude 6.5) distributed over the entire celestial sphere. To these fixed points of light, we can add the Moon, a total of six naked-eye planets wandering among the fixed stars, frequent meteors (shooting stars) that quickly dart across the sky, and the occasional passage of comets in their eccentric orbits around the Sun, which add yet another dynamic aspect to the night sky.

Of the earliest cosmologies, Aristotle's classic work *On the Heavens* stands out as the most influential, dominating Western philosophy from the time it was written (350 BC) up through the Middle Ages. Aristotle (384–322 BC) divided the celestial realm into a number of nested crystalline spheres to which the Moon, Sun, and planets were attached. The motions of these bodies were coupled to one another, and to the outermost sphere—the realm of the fixed stars—by a number of intermediate layers. A schematic illustration of Aristotle's Universe is depicted in Figure 1.1.

The whole assembly was animated by a "prime mover" who acted solely in the outermost sphere, which was thought to be immutable and eternal. As Aristotle put it in *On the Heavens*: "In the whole range of time past, so far as our inherited records reach, no change appears to have taken place either in the whole scheme of the outermost heaven or in any of its proper parts." In other words, although celestial objects in the more Earthly realms moved and varied (e.g., the Moon, the planets, meteors, and comets), the stars were strictly fixed and unchanging in his model.

One doesn't have to ponder long to realize that had Aristotle been correct, the study of astronomy would be quite limited. Fortunately, before long it became clear that the realm of the stars was far from the static one that Aristotle had imagined. In the Far East, and worlds away, ancient Chinese sky-watchers were already documenting the appearance of "guest stars" as early as 200 BC (Ho 1962). Figure 1.2 illustrates a page from the Chinese manuscript Lidai mingchen zouyi

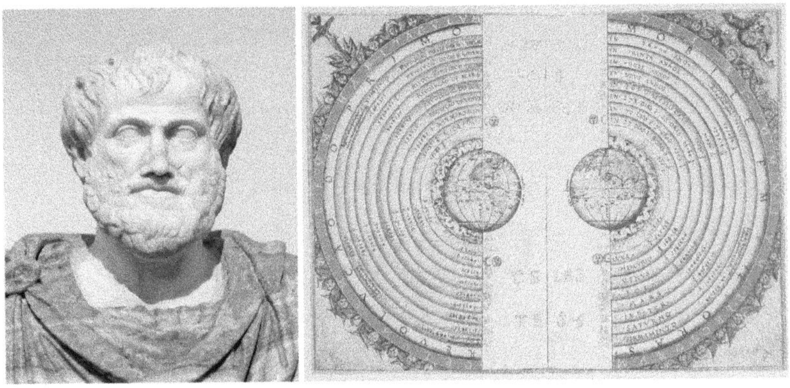

(a) (b)

Figure 1.1. (a) Aristotle (Roman copy in marble of a Greek bronze by Lysippos, c. 330 BC). (b) Bartolomeu Velho's 1561 depiction of a geocentric Aristotelian Universe. (Credit: Wikipedia Commons.)

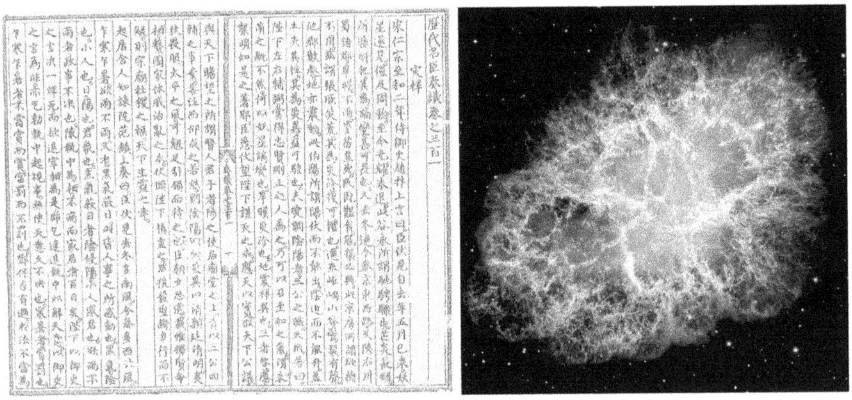

(a) (b)

Figure 1.2. A page from early Chinese records, the Lidai mingchen zouyi from the year 1414. The highlighted passages describe the supernova of 1054 that ultimately produced the supernova remnant known as the Crab Nebula that we see today (right, courtesy NASA). The translation reads: 2nd year of the Zhihe reign period of Emperor Renzong of Song [1055]; Attendant Censor Zhao Bian submitted a letter saying: "Your servant considers that, since the 5th month of last year [when] the baleful star appeared, a full year has passed and until now its brilliance has not faded [lit. "retreated"]. This is what Gu Yong meant by 'its rapid movement, the variations in the length of its flaming rays, and the [asterisms] on which it has trespassed successively,' as a censorious anomaly it is greatly to be feared." (Pankenier 2006). (Credit: Wikipedia Commons.)

(circa 1414), describing the "guest star" of 1054, which was later determined to have produced what we now refer to as the Crab Nebula.

In contrast to the East, the medieval Western world seemed surprisingly disinterested in monitoring variation in the night sky. As is well known, it is remarkable that the supernovae of 1006 (the brightest supernova ever recorded) and 1054, which were estimated to have reached apparent magnitudes of −7.5 and −6, respectively, produced little attention in the West. We have to wait until the late

Renaissance or Early Modern period when two of the most spectacular "guest stars," now known by their Latin name, Novae Stella, or "new stars," were observed (within a span of just 32 years!): Tycho's nova of 1572 and Kepler's nova of 1604. Both of these objects were later recognized as supernovae, and were the most recent two such events seen in the Milky Way. A comprehensive discussion of the early history of nova observations can be found in the review by Duerbeck (2008).

As remarked by Warner (1995) in his classic book *The Cataclysmic Variable Stars,* despite the fact that recent observations show that roughly one nova reaches second magnitude or brighter every decade or so, only a single "guest star," Nova Vulpecula 1670 (CK Vul)[1] was recorded between 1604 and the time when Nova Ophuiuchi 1848 (V841 Oph) was discovered.[2] Apparently, the seventeenth and eighteenth centuries did not produce active communities of committed sky-watchers!

Interest in the transient sky picked up noticeably during the second half of the nineteenth century, thanks in large part to the application of photography to astronomy, and the number of novae discovered in the Milky Way skyrocketed. By 1900 a total of some 25 nova candidates had been reported. Table 1.1 summarizes 15 of the most secure nova identifications from that period. Despite the proliferation of nova discoveries, the true nature of novae remained obscure. There was no serious model to explain the phenomena, and it would be another century before the cause of nova eruptions would be finally established (see Chapter 4).

1.1 The Nebulae

The development of the telescope not only revolutionized the study of variable stars, but it opened up a new dimension to the study of the heavens: exploration of the enigmatic "nebulae." The nebulae, diffuse sources of light in the night sky, had been recognized since antiquity. Claudius Ptolemaeus (Ptolemy) (AD 100–170) described several entries in his second century AD star catalog as being "nebulous," many of which are now known to be star clusters. The great nebula in Andromeda, which at an integrated apparent magnitude of $\sim$4 is visible to the unaided eye, was recognized as early as AD 964 by Abu I-Husain al-Sufi in his book "*On the Constellations of the Fixed Stars*" (Kunitzsch 1987). The number and variety of nebulous sources expanded considerably in the eighteenth and nineteenth century due in large part to the telescopic observations of Edmund Halley, Charles Messier, and especially the Herschels (William, his sister Caroline, and later William's son, John). The catalogs compiled over years by the Herschels, and augmented by John Louis Emil Dreyer, were eventually published as the *New General Catalog* (NGC) in 1888 (Dreyer 1888).

[1] Variable star designations are made in the order of discovery and begin with the letter R followed by the name of the constellation. The sequence continues through Z, then RR through RZ, SS to SZ, and so on to ZZ. The sequence then shifts to the beginning of the alphabet AA to AZ, BB to BZ, ending at QZ (the letter J is always omitted). This sequence contains 334 names, after which the variables are named V335, V336, etc. The nomenclature was devised by Friedrich Wilhelm August Argelander (1799–1875) in the nineteenth century.

[2] The identification of CK Vul as a classical nova is now questioned (Hajduk et al. 2013), however, Warner's comment remains accurate as Nova Sagittae 1783 (WY Sge) was not included in his discussion.

Table 1.1. Well Established 19th Century Novae

Nova	Name	Peak mag	Type	Discoverer	Reference
N Sgr 1900	HS Sgr	11.5 p	NB:	I. E. Woods	Woods (1927)
N Sgr 1900	AT Sgr	11.0 p	NA:	H. S. Leavitt	Pickering (1904)
N Sgr 1899	V1016 Sgr	8.5 p	NB:	A. J. Cannon	Pickering (1910)
N Aql 1899	V606 Aql	6.7 p	NA	W. Fleming	Fleming (1900)
N Oph 1898	RS Oph	4.3 v	NR	W. Fleming	Fleming (1901)
N Sgr 1898	V1059 Sgr	4.9 p	NA	W. Fleming	Fleming (1899)
N Car 1895	RS Car	7.0 p	N	W. Fleming	Fleming (1895)
N Nor 1893	IL Nor	7.0 p	NB	W. Fleming	Fleming (1893)
N Aur 1891	T Aur	4.2 p	NB	Th. D. Andersen	Copeland (1892a, 1892b)
N Per 1887	V Per	9.2 p	N	W. Fleming	Fleming (1891)
N Cyg 1876	Q Cyg	3.0 v	NA	J. Schmidt	Schmidt (1877)
N CrB 1866	T CrB	2.0 p	NR	J. Birmingham, W. Barker	Lynn (1866a, 1866b); Hind et al. (1866)
N Sco 1863	U Sco	8.8 v	NR	N. R. Pogson	Pogson (1908)
N Sco 1860	T Sco	7 v	N:	E. Luther, A. Auwers	Luther & Auwers (1860)
N Oph 1848	V841 Oph	4.2 v	NB	J. R. Hind	Hind (1848a, 1848b)

Note. N: Nova of undetermined speed class; NA: fast nova; NB: slow nova; NR: recurrent nova.

Dryer later supplemented the NGC catalog with additional nebulae in his *Index Catalog*, IC (Dreyer 1895, 1908).

By the mid-nineteenth century several thousand nebulae had been identified, but the true nature of these objects remained a mystery. William Herschel (1738–1822) had originally believed that the nebulae were simply unresolved clusters of stars. However, in 1790 he noticed that one roughly circular nebula (now known as NGC 1514, and one of the objects that he referred to as a "Planetary Nebula") was centered on a 9th magnitude star. He correctly associated the nebulosity with the central star, which caused him to re-think his conjecture that all nebulae were unresolved star clusters.

A major step forward in our understanding of the nature of nebulae was made possible in the 1840s thanks primarily to advances in telescope technology; in particular, to the construction of a giant 72 inch reflector, the "Leviathan" at Parsonstown by William Parsons (1800–1867), aka Lord Rosse, the 3rd Earl of Rosse (see Figure 1.3). The increased aperture and quality of the mirror figure compared with previous instruments dramatically improved the level of detail one could discern in the spatial structure of nebulae. Of the many discoveries made with the 72 inch telescope perhaps the most consequential was the discovery by Rosse of the spiral structure of M51 (see Figure 1.4). Over the next several decades, aided by the increasing application of photography to the study of nebulae, it became apparent that a spiral structure was not uncommon among the nebulae.

Just two decades after the discovery of spiral nebulae, a new and powerful tool was brought to bear in the attempt to understand the nature of the nebulae: the spectroscopic analysis of nebular light by William Huggins (1824–1910). Huggins

Figure 1.3. (Left) The Leviathan of Parsonstown was a 72 inch reflecting telescope used by William Parsons, 3rd Earl of Rosse, in his study of spiral nebulae. It was the largest telescope in the world until the completion of the Mount Wilson 100 inch "Hooker" reflector in 1917 (undated photo). (Right) Portrait of Lord Rosse by Stephen Catterson Smith. (Credit: Wikipedia Commons.)

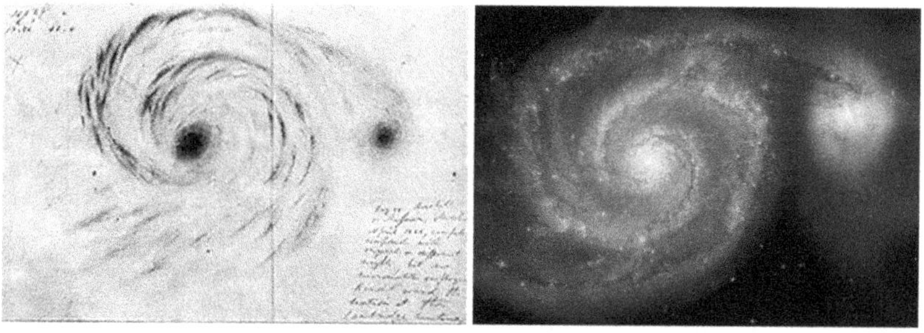

Figure 1.4. A Sketch of the spiral structure of M51 by Lord Rosse (left) compared with a modern image from the *Hubble Space Telescope* (right, Credit: ESA/Hubble). (The Rosse sketch is courtesy of the Birr Castle Archives from a folio scrapbook entitled "Astronomical Drawings of Lord Rosse," Rosse Papers, L/3/2.)

(shown in Figure 1.5) conducted spectroscopic observations that revealed many nebulae to have bright-line (emission) spectra, therefore suggesting that they were gaseous in nature (Huggins & Miller 1864). Quoting from Huggins & Miller (1864) concerning his spectroscopic analysis of NGC 6543, the "Cat's Eye" planetary nebula in Draco: "I then found that the light of this nebula, unlike any other ex-terrestrial light which had yet been subjected by me to prismatic analysis, was not composed of light of different refrangibilities, and therefore could not form a spectrum. A great part of the light from this nebula is monochromatic, and after passing through prisms remains concentrated in a bright line occupying in the instrument the position of that part of the spectrum to which its light corresponds in refrangibility."

Thus, Huggins had established that a substantial fraction (roughly a third) of the nebulae were in fact *not* unresolved systems of stars as Herschel had long assumed,

Figure 1.5. Portrait of Sir William Huggins by John Collier in 1905. (Credit: Wikipedia Commons.)

but were bright (emission) line objects such as the Orion nebula. Nebulae that displayed strong, bright lines at 4959 and 5007 Å (which wouldn't be understood for another 75 years as forbidden lines of doubly ionized oxygen) were often the planetary nebulae found by Herschel, and based on their color, were referred to by Huggins as the "Green Nebulae." The remaining nebulae apparently characterized by continuous spectra were referred to as the "White Nebulae." Once photographic images of the White Nebulae became common, it became clear that most of these objects were in fact Spiral Nebulae. It is worth noting that Huggins has the distinction of being the first to obtain a spectrum of a nova (Huggins & Miller 1866), specifically the recurrent nova T Corona Borealis discussed further in Chapter 4.

Observations of nebulae compiled in the *General Catalogue of Nebulae and Clusters of Stars* published by Sir John Herschel in 1864 confirmed what had been suspected for some time, namely that the nebulae tended to avoid portions of the sky with a high density of stars, i.e., the Milky Way (or *Via Lactea*; Abbe 1867). The region populated by the nebulae was subsequently described by Proctor (1869) as the "Great Vacant Zone," or the "Zone of Dispersion." Thus, the complementary spatial distributions suggested that nebulae and stars were not necessarily associated with one another as Herschel had originally believed.

As the nineteenth century was drawing to a close, the number of nebulae was increasing dramatically as a result of the implementation of photographic surveys. The observations of James E. Keeler (1857–1900) using the Lick Observatory's Crossley reflector suggested that the number of spiral nebulae distributed over the whole sky accessible to the Crossley telescope would number of order 120,000 (Keeler 1899). Whatever their nature, it was already clear that the spiral nebulae represented a major constituent of the visible Universe.

1.2 Novae in Nebulae!

In 1885 a "nova," S Andromedae (S And), was observed close to the center of NGC 224 (M31), the great nebula in Andromeda. Official credit for the discovery of S And on 1885 September 20 would go to Hartwig (1885); however, the object was apparently observed in the preceding days by a number of amateur astronomers (see de Vaucouleurs & Corwin 1985 for a detailed account of the discovery). Little did Hartwig know that this object was to play a pivotal role in the debate on the nature of the spiral nebulae that would unfold over the next 40 years. It wouldn't be until the 1930s that the true nature of S And would be revealed as a supernova in M31. The remnant of explosion was finally recovered more than a century after S And erupted (see Figure 1.6). It would be another decade before the next supernova was discovered in the field of a spiral nebula.

During the latter part of the nineteenth century, Edward C. Pickering (1846–1919), director of the Harvard College Observatory (HCO) from 1877 to 1921, employed a staff of mostly female assistants (colloquially known as "computers"), who were trained to do routine data processing tasks, e.g., measuring the positions, brightnesses, and colors of stars. While some of the women had formal training in the sciences, not all of them did. One notable example was Williamina Fleming née Stevens (1857–1911). Williamina Stevens was born in Scotland in 1857, but immigrated to the United States with her husband, James Fleming, when she was 21. Shortly after arriving in Boston, Mrs. Fleming, who was pregnant at the time, was abandoned by her husband. Although she had been previously employed as a teacher in Scotland, times were now more desperate, and in 1879 Fleming applied

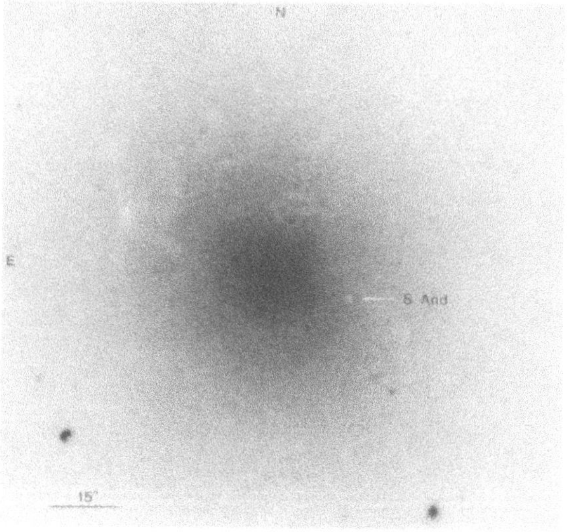

Figure 1.6. The remnant of S Andromadae as revealed by isolating Fe I emission commonly observed in Type Ia supernovae. Note the proximity of S And to the nucleus of M31. (Reproduced from Fesen et al. 1989. © 1989. The American Astronomical Society. All rights reserved.)

for and was offered a position as E. C. Pickering's maid. At the time, Pickering employed several men as assistants at HCO who primarily were charged with the tedious and repetitive work of examining and classifying the spectra of stars. Pickering was not particularly impressed by the quality of work provided by his assistants, and was said to have remarked that even his "Scottish maid" could do better. Apparently Pickering was not being facetious, and in 1881 he hired Williamina Fleming as a "computer" at HCO. Shortly thereafter, Fleming was put in charge of hiring additional women to help with the work, and the group known as "Pickering's Harem" was born (see Figure 1.7).

Over the next several years, Fleming would go on to hire such notable women as Annie Jump Cannon and Henrietta Levitt, two women who would later make significant contributions to astronomy in the areas of stellar classification and the Cepheid period–luminosity relation, respectively. Fleming herself made many important contributions, including classifying the majority of stellar spectra used in the Henry Draper Catalog (published in 1890). She would also make many notable discoveries, including the first recognition of the "Horsehead Nebula" in 1888 on a plate that had been taken years earlier by Pickering. Her discoveries also included over 200 variable stars, several of them novae (see Table 1.1), one of which was the second "nova," after S And, to be associated with a spiral nebula. That apparent nova, Z Centauri, erupted near the center of NGC 5253, and reached

Figure 1.7. Staff at work at the Harvard College Observatory Computer room in 1891. Known informally as "Pickering's Harem": the women were engaged in a variety of data analysis tasks. HCO director E. C. Pickering looks on (left), with Williamina Fleming (standing center). (Image courtesy of the Harvard University Archives (HUV 1210 (9-4), olvwork289689).)

magnitude 7.2 sometime between 1895 June 14 and July 8. It was subsequently discovered by Mrs. Fleming while examining plates on December 12 of that year (Pickering & Fleming 1896). As it happened, novae were to play a major role in the birth and development of extragalactic astronomy.

References

Abbe, C. 1867, MNRAS, 27, 257
Copeland, R. 1892a, AN, 129, 59
Copeland, R. 1892b, AN, 129, 75
de Vaucouleurs, G., & Corwin, H. G. Jr. 1985, ApJ, 295, 287
Dreyer, J. L. E. 1888, MmRAS, 49, 1
Dreyer, J. L. E. 1895, MmRAS, 51, 185
Dreyer, J. L. E. 1908, MmRAS, 59, 105
Duerbeck, H. W. 2008, Classical Novae, ed. M. F. Bode, & A. Evans (2nd ed.; Cambridge: Cambridge University Press)
Fesen, R. A., Hamilton, A. J. S., & Saken, J. M. 1989, ApJ, 341, L55
Fleming, M. 1891, AN, 126, 163
Fleming, M. 1893, AN, 134, 59
Fleming, M. 1895, AN, 139, 119
Fleming, M. 1899, AN, 149, 13
Fleming, M. 1900, AN, 153, 59
Fleming, M. 1901, AN, 154, 425
Hajduk, M., van Hoof, P. A. M., & Zijlstra, A. A. 2013, MNRAS, 432, 167
Hartwig, E. 1885, AN, 112, 358
Hind, J. R. 1848a, AN, 27, 191
Hind, J. R. 1848b, MNRAS, 8, 146
Hind, J. R., Bird, F., & Lynn, W. T. 1866, AReg, 4, 229
Ho, P.-Y. 1962, VA, 5, 127
Huggins, W., & Miller, W. A. 1864, RSPT, 154, 437
Huggins, W., & Miller, W. 1866, RSPT, 15, 146
Keeler, J. 1899, AN, 151, 1
Kunitzsch, P. 1987, Msngr, 49, 42
Lynn, W. J. 1866a, AN, 67, 197
Lynn, W. J. 1866b, AReg, 4, 310
Luther, E., & Auwers, A. 1860, AN, 53, 293
Pankenier, D. W. 2006, JAHH, 9, 77
Pickering, E. C., & Fleming, W. P. 1896, ApJ, 3, 162
Pickering, E. C. 1904, AN, 167, 165
Pickering, E. C. 1910, AN, 186, 319
Pogson, N. R. 1908, MmRAS, 58, 90
Proctor, R. A. 1869, MNRAS, 29, 337
Schmidt, J. F. J. 1877, AN, 89, 9
Warner, B. 1995, The Cataclysmic Variable Stars (Cambridge: Cambridge University Press)
Woods, I. E. 1927, BHarO, 851

Chapter 2

Novae and the Nature of Spiral Nebulae

By the close of the nineteenth century the debate over whether the spiral nebulae were "island universes" comparable to the Milky Way and composed of multitudes of stars as speculated in 1755 by Immanual Kant (1724–1804), or whether they were composed of a gaseous nebulous fluid within the Milky Way as proposed by Pierre Laplace (1749–1827) in his book *Exposition du système du monde (Exposition of a World System)* published in 1796, had not yet begun in earnest. Nevertheless, the preponderance of evidence had tipped decidedly in favor of the latter hypothesis. In particular, the observations of Huggins demonstrating the gaseous nature of the Green (Planetary) Nebulae, the failure of early photographic observations to resolve the spiral nebulae into stars, along with the conjecture dating back to Laplace that nebulae were associated with the formation of stars, seemed to provide strong arguments against the island universe hypothesis.

The two apparent novae observed in NGC 224 and NGC 5253 also seemed problematic for the notion that spiral nebulae were vast conglomerations of stars. How could the brightness of the two novae rival the integrated light of a stellar system like the Milky Way? After all, the nova T Sco (Nova Sco 1860) observed in the globular cluster M80 seemed to make it clear that nova luminosities were far lower than that of the Milky Way! Famed historian of astronomy Agnes Mary Clerke (1842–1907; Clerke 1905) summed up the situation at the end of the nineteenth century thusly: "The question of whether nebulae are external galaxies hardly any longer needs discussion. It has been answered by the progress of research. No competent thinker with the whole of the available evidence before him can now, it is safe to say, maintain any single nebula to be a star system coordinate rank with the Milky Way." There is a lesson in this that transcends the ages: be wary of premature pronouncements of "Settled Science." The distinction between novae and supernovae was not yet appreciated. The next 25 years would be revolutionary.

During the two decades that followed the discovery of Z Cen in 1895, a total of ~20 new novae were discovered, with the rate remaining constant at about one per

doi:10.1088/2514-3433/ab2c63ch2

year on average. Among these were the spectacular nova of 1901 in Perseus, GK Per, which reached $m \sim 0.2$, briefly becoming one of the 10 brightest stars in the sky. In addition, the second outburst of U Sco (Nova Sco 1863) was seen in 1906, making it the first of the "recurrent novae" (RNe) to be so identified. During this period, however, no new novae were discovered to be associated with spiral nebulae. In retrospect, this might appear somewhat surprising given the considerable interest in the nature of the spiral nebulae these objects were imaged quite frequently. Apparently, despite the discovery of S And and Z Cen in the latter years of the nineteenth century, there was no concerted follow-up effort to search for novae in the spiral systems.

2.1 Proper Motions and Spiral Nebulae

Beginning in 1912, Vesto Slipher (1875–1969) working at Lowell Observatory began a program of measuring the Doppler shifts of spiral nebulae.[1] Slipher's first target was the Andromeda nebula, which he found to be approaching the Sun with a velocity of 300 km s^{-1}. At the time, this was the largest radial velocity measured for any celestial object (Slipher 1914). He would go on to measure radial velocities for several additional spiral nebulae, realizing that such data might provide further insights into the nature of these systems. Slipher concluded his 1914 paper on the radial velocity of M31 with the following comment: "Thus extension of this work to other objects promises results of fundamental importance, but the faintness of the spectra makes the work heavy and the accumulation of results slow."

Shortly after Slipher's discovery of the rather large radial velocities shown by the spiral nebulae, the Lick Observatory's Heber Doust Curtis (1872–1942), who would later become the leading proponent for the extra-Galactic nature of the spiral nebulae, realized that the average distances to the spiral nebulae could be deduced if the proper motions of the nebulae could be measured (Curtis 1915). The ability to detect proper motions reliably requires high quality (sharp) photographic images well separated in time. For his study, Curtis made use of the plates available from the Lick Observatory's Crossley reflector—the same instrument that Keeler had employed to make his survey of spiral nebulae in the late nineteenth century. All told, Curtis analyzed the positions of 66 spiral nebulae, in some cases using the position of the nucleus itself, and in other cases the positions of various stellar condensations. His measurements revealed proper motions averaging 0.033 arcsec yr^{-1}. For the same sample, velocity measurements by Slipher averaged 400 km s^{-1}. Under the assumption that the motions were truly random, Curtis was able to derive average distances to the nebulae. Curtis himself appeared to be skeptical of the results, concluding his paper by stating "… the average distance of the spirals is of the order of ten thousand light-years, a distance which many will regard as too small." Future observations, of course, would reveal the proper motion measurements to be spurious.

[1] A description of the technical details of his observations is given by Thompson (2011).

While measuring radial velocities for additional spiral nebulae, Slipher noticed a wavelength shift across the major axis of NGC 5494 (see the discussion by Brèmond 2009). The discovery that spiral nebulae were rotating would lead directly to attempts to measure proper motions internal to the nebulae. Unfortunately, over the ensuing decade that pursuit would lead efforts to establish the true nature of the spiral nebulae far astray. The discovery by Slipher that the spiral nebulae were rotating led to an interest in looking, not just for proper motions of the nebulae themselves, but for proper motions *internal* to the nebulae themselves.

In 1915 December, George Willis Ritchey (1864–1945),[2] an optician by training and an associate of Hale working at the Mount Wilson Observatories, asked Adriaan van Maanen (1884–1946), a Dutch astronomer who had received his doctorate from the University of Utrech in 1911, to search for proper motions in two archival photographs he had taken of the magnificent face-on spiral nebula M101. Van Maanen was the perfect man for this challenging task having focused his graduate studies on proper motion measurements of star clusters—work that resulted in his thesis entitled "The Proper Motions of 1418 Stars in and near the clusters h and χ Persei." In an historically significant paper published in the Astrophysical Journal, van Maanen analyzed five plates of M101: three taken by Ritchey himself, one taken in 1899 by Keeler, then Director of Lick Observatory, and the fifth by Charles D. Perrine of the Lick Observatory in 1908. The result of van Maanen's study was the discovery of internal proper motions that averaged 0.022 arcsec per year at a mean radius of 5 arcmin from the center of the nebula (see Figure 2.1). Taken at face-value, this suggested a rotation period of order 85,000 years.

2.2 Novae and Nebulae

The study of novae in spiral nebulae eventually experienced a renaissance in 1917, when in July of that year, Ritchey serendipitously discovered a nova in the spiral nebula NGC 6946 while examining images of spiral nebulae for internal proper motions (Ritchey 1917a, 1917b). It is noteworthy that this nova was the first to be discovered in more than five years, the last being the Galactic[3] nova DN Gem in March of 1912.

Ritchey's discovery of the nova in NGC 6946 spurred him to go back and re-examine all earlier plates taken with the Mount Wilson 60 inch reflector since its completion in 1908. A total of six novae or variable objects were identified in four different nebulae: one in NGC3031 (M81), one in NGC2403, two in NGC5457 (M101), and two in NGC224 (the Andromeda nebula, M31). Of these, it is likely that only two or three were actually novae (one in NGC 2403 and possibly two in Andromeda), with the others, which were seen over a period of several years, likely being long-period variable stars. The two nova candidates identified by Ritchey in

[2] Ritchey is perhaps best known for the development, along with Henri Chrétien, of the Ritchey–Cretien (RC) reflecting telescope design.

[3] The terms "The Galaxy" and "Galactic" refer to the Milky Way galaxy and objects related to the Milky Way, respectively.

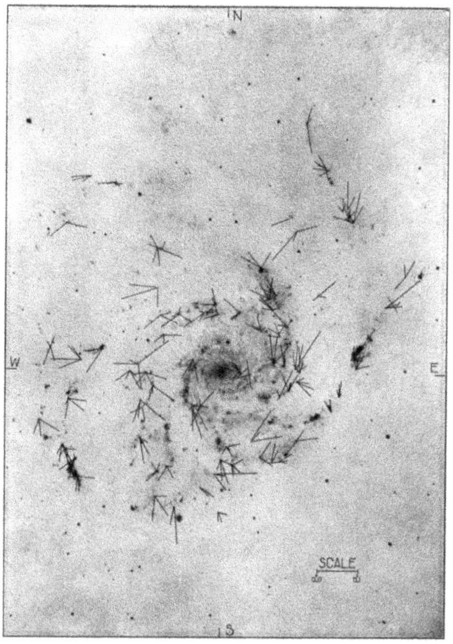

Figure 2.1. Adriaan van Maanen's measurements of the internal proper motions in the face-on "pinwheel" galaxy, M101. (Reproduced from van Maanen 1916. © 1916. The American Astronomical Society. All rights reserved.)

Andromeda were discovered contemporaneously in 1909 September during the first year of operation of the 60 inch telescope. A reproduction of one of Ritchey's plates taken on 1909 September 13 is shown in Figure 2.2. The bottom panel shows an enlargement of the region approximately 3′ west and 2′ south of the center of the nebula where the nova candidates were found. The southernmost nova, now recognized as M31N 1909-09b[4] is clearly visible on this plate. The other nova candidate, M31N 1909-09c, not yet near peak brightness, is barely visible. According to Ritchey's initial photometric measurements, 1909-09b reached a peak brightness of $m_{pg} = 16.3$ on September 15, with 1909-09c reaching $m_{pg} = 17.0$ a day later. A re-analysis by Hubble (1929) of the original plates found peak magnitudes of 16.7 and 17.1 for 1909-09b and 1909-09c, respectively. The photometric evolution of 1909-09b was typical of a nova, taking about three weeks to fade by two magnitudes. On the other hand, 1909-09c faded quite slowly, and was still visible ~2 mag below peak on November 7.

The discovery of the six nova candidates by Ritchey on the Mount Wilson 60 inch plates led to a more widespread search for novae on archival images of the spiral nebulae. Within a matter of a few months, Harlow Shapley (1885–1972; Shapley 1917), who would later go on to champion the Galactic nature of the nebulae,

[4] The Central Bureau for Astronomical Telegrams (CBAT) designation for M31 novae is M31N YYYY-MMa, where YYYY is the 4-digit year, MM the 2-digit month, and "a" is a lower-case letter representing the order of discovery within the month.

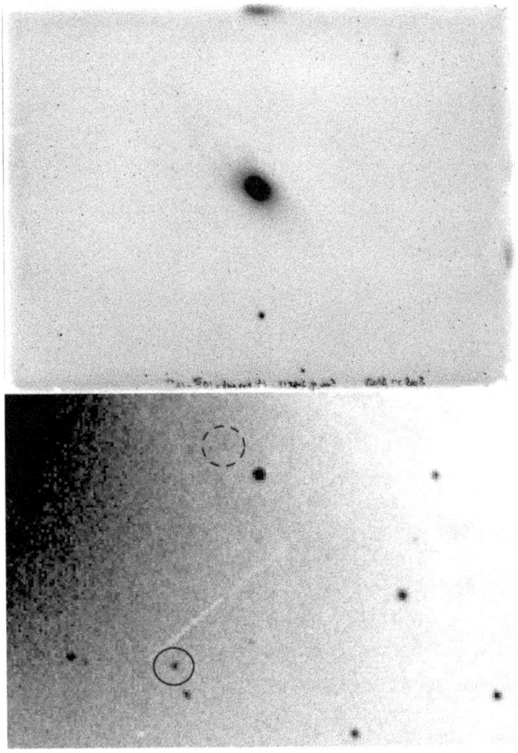

Figure 2.2. (Top) A Mount Wilson 60 inch plate taken by George Ritchey on the night of 1909 September 13. (Bottom) The first bona fide classical nova ever discovered in M31, M31N 1909-09b, can be seen on a scan of the Ritchey plate (solid circle). The position of a more questionable nova candidate, M31N 1909-09c, is indicated by the dashed circle. (Images courtesy of the Carnegie Institution for Science.)

compiled a list of 11 novae associated with various spiral systems. One of these objects was a variable source in the field of NGC 3147. The object had been identified by Mrs. Dorothea Roberts in 1913 on a plate taken nine years earlier by her late husband Isaac Roberts, a pioneer in astrophotography and member of the British Astronomical Association. Years later subsequent outbursts would be observed, and the object was determined to be a foreground dwarf nova,[5] CP Dra, in the Milky Way (see Shears 2011). Seven other objects, including S And and Z Cen were later identified as supernovae, with the remaining three objects, Ritchey's two 1909 M31 sources plus a nova discovered by Shapley in 1917 September, revealed to be common novae associated with M31. Ritchey (1917c) would subsequently find another M31 nova on 1917 October 16.

First Curtis (1917) and then Shapley (1917) went on to compare the novae in spiral nebulae with field (Galactic) novae. The average magnitude of the Galactic

[5] Dwarf novae are discussed in Chapter 4.

novae known at the time was +5.5,[6] while that of novae seen in spiral nebulae was +14. Assuming that the novae were similar (which they weren't as most of the novae in spirals were later to be identified as supernovae), Shapley found that the spiral nebulae must be ~50 times farther away than the average distance of the Milky Way novae. Curtis had come to a similar conclusion a few months earlier after assuming a slightly brighter average apparent magnitude of +5 for Galactic novae (to account for likely missing the peak magnitudes). For the spiral nebula sample he made a slight correction for the presumed observational bias towards the brightest novae, and adopted a somewhat fainter average magnitude of +15. This 10 magnitude difference led Curtis to conclude that the spiral nebula novae were 100 times more distant than the average distance of the Galactic sample, and thus clearly outside the Milky Way.

Shapley went on to compare the angular diameters of the spiral nebulae with his estimate of the absolute size of the Milky Way (a minimum of 50,000 light years), determining that the novae must have a maximum absolute magnitude of −7, and in the case of S And, a maximum of −15. This tremendous luminosity seemed preposterous to Shapley, who argued that "This remarkable result must inevitably follow if spiral nebulae are considered external galactic systems comparable with our own in size and constituency." In the same paper Shapley marshalled additional support for his view that the spiral nebulae lay within the confines of the Milky Way by pointing out that the purported 85,000 year rotation period for M101 measured by van Maanen (1916) would imply a rotational velocity that exceeded the speed of light *if* M101 was a stellar system comparable in size to the Milky Way.

To recap, as 1917 was drawing to a close, there were two principal arguments against the island universe model: the seemingly implausible luminosity of S And, and the superluminal rotational velocity of M101 implied by van Maanen's internal proper motion measurements. Despite these arguments, which were considered quite compelling at the time, Curtis remained firmly in the island universe camp.

2.3 The Great Debate

At the dawn of the new decade the debate concerning the nature of the spiral nebulae was no closer to a resolution, with the astronomical community divided on the question of whether the spiral nebulae were "local," i.e., within the Milky Way, or separate "island universes." On 1920 April 26 a pair of lectures were delivered at the annual meeting of the National Academy of Sciences in Washington DC by proponents of each view—Harlow Shapley making the case for the former interpretation, and Heber Curtis arguing for the latter. The resulting lecture and discussion became known as "The Great Debate." The definitive account of what transpired has been chronicled by the historian of science Michael A. Hoskin (1976).

The genesis of the Great Debate can be attributed to the renowned astronomer and visionary, George Ellery Hale (1868–1938), a towering figure in astronomy at

[6] See the list of known novae in Fleming & Pickering (1912, p. 169) omitting the spiral nebula (super)novae S And and Z Cen.

the dawn of the last century.[7] The son of a wealthy Chicago businessman, Hale was the driving force behind the planning and construction of many of the world's most powerful telescopes, including the 40 inch refractor at Yerkes Observatory, the 60 inch and 100 inch reflectors at Mount Wilson Observatory (which he founded with support from the Carnegie Institution of Washington), and eventually the iconic 200 inch reflector at Palomar Observatory that was completed in 1949, a little more that a decade after his death. Hale was primarily a solar astronomer, who is credited with the invention of the spectroheliograph while an undergraduate at the Massachusetts Institute of Technology. He would later employ a modified spectroheliograph to establish through the Zeeman effect that sunspots were magnetic. Hale's other notable achievements included his role from 1895 to 1902 as the first editor of the Astrophysical Journal (along with Lick Observatory's James E. Keeler), and his appointment to the National Academy of Sciences (NAS) in 1902.

Hale was quite influential within the NAS, and had set up an endowment for invited lectures at the annual NAS meetings. In late 1919 Hale approached Charles G. Abbot, home secretary of the NAS, and suggested, in honor of his father, that there be a William Ellery Hale lecture at the 1920 meeting of the NAS. Hale preferred that the lecture focus on one of two topics of widespread interest at the time: General Relativity or the Scale of the Universe, and that the format be that of a debate rather than a traditional lecture. In early 1920 Abbot wrote to Hale concerning his thoughts on the debate (Hoskin 1976): "You mentioned the possibility of a sort of debate, either on the subject of the island universe or of relativity. From the way the English are rushing relativity in Nature and elsewhere it looks as if the subject would be done to death long before the meeting of the Academy, and perhaps your first proposal to try to get Campbell and Shapley to discuss the island universe would be more interesting. I have a sort of fear, however, that the people care so little about island universes, notwithstanding their vast extent, that unless the speakers took pains to make the subject very engaging the thing would fall flat ··· Are there not other subjects—the cause of glacial periods, or some zoological or biological subject—which might make an interesting debate?"

Although Hale apparently favored relativity, Abbot was worried that the subject would be too esoteric for the intended audience, and pushed for a debate on the scale of the Universe. Abbot eventually prevailed and after further consideration it was decided that Heber Curtis, rather than the more senior W. W. Campbell (director of Lick Observatory at the time), would be the more appropriate advocate for the island universe position. Invitations were sent to Curtis and Shapley, both of whom accepted; although there was considerable negotiation concerning the format of the event. Curtis, apparently looking forward to a "friendly scrap" was happy with a debate format with ample time for each man to make their case. Shapley, on the other hand, who at the time was angling to be the successor of Edward C. Pickering as director of Harvard College Observatory, was much more cautious. He strongly preferred that the event take the form of a "discussion" consisting of two separate

[7] An excellent biographical memoir of Hale was written for the National Academy of Sciences by Walter S. Adams (1939).

talks of just 35 minutes in length. When the dust had settled, a format consisting of two separate presentations, 40 minutes in length with Shapley going first, was finally agreed to and adopted.

Given the structure of the event, each man had considerable latitude in how they approached the subject in their presentations. Shapley felt he was on firmer ground in discussing the dimensions of the Milky Way rather than address the nature of the spiral nebula directly. His presentation, qualitatively described how Cepheid variables and B stars could be used to determine distances to the globular clusters, the most distant of which were assumed to reflect the maximum linear dimension of the Milky Way's disk. Shapley's conclusion, based on the spatial distribution and distances of the globular clusters was that the Sun was some 60,000 light years from the center of the Milky Way, which had an overall diameter of order a staggering 300,000 light years.[8]

Curtis, on the other hand, focused his remarks on the spiral nebulae, pointing out that their spectra were indistinguishable from those of clusters of stars and of the integrated light of the Milky Way generally, and not at all like the known gaseous nebulae, which were characterized by prominent emission lines. He went on to argue that the "zone of avoidance" (the apparent dearth of spiral nebulae near the plane of the Milky Way), could be explained by obscuring material in the plane, as was typically seen in elongated (edge-on) spiral nebulae. Without directly countering Shapley's arguments for a 300,000 light year diameter for the Milky Way, Curtis did take a swipe by pointing out how Shapley's value far exceeded previous determinations.

When the "debate" was over, the general consensus was that Curtis had prevailed. He was the more articulate speaker, and had presented his case in a more quantitative manner. In correspondence between the two men after the fact, Shapley offered that perhaps his presentation may have been too "popular" in character compared with Curtis' more "technical" presentation. Curtis seemed to concur, and admitted to Shapley that he had been caught off guard by the "general character" of Shapley's remarks. Sir Arthur Eddington, the renowned British astrophysicist, who along with Shapley was in the running for the directorship of Harvard College Observatory, took the opportunity to take a shot at Shapley's performance at the debate, telling Hale that if Shapley came to Harvard as (the accomplished stellar astronomer) Henry Norris Russell's (1877–1957) "second" (in command) he should offer a lecture course for this "cultivates the gift of gab, which he needs." In any event, Shapley's performance was well enough received that it didn't damage his quest to become the next director at Harvard, a position to which he was appointed in 1921 October.[9]

[8] The Cepheid zero-point calibration was made through observations of relatively nearby (disk) objects, which, as Walter Baade would show some two decades later, are more luminous than their Cepheid counterparts in the globular clusters. As a result, Shapley overestimated the luminosities of the cluster Cepheids leading him to overestimate their distances. Interestingly, this is the same problem (the other way around) that would lead Hubble to underestimate the distance to the Andromeda nebulae some five years later.

[9] A excellent account of the search for the director position can be found in the account by Gingrich (1988).

2.4 The Birth of Extragalactic Astronomy

In the years following the realization that novae were relatively common in spiral nebulae, up through the start of Hubble's systematic nova survey in M31 described below, interest in novae among professional and amateur astronomers alike was quite strong. A cursory examination of the literature between 1917 and 1923 from the Astronomical Data System (ADS) reveals that nearly 500 articles were published on novae or "temporary stars," which eclipsed the number of papers published on the general topic of "nebulae."[10] For comparison, despite their role in enabling Shapley to measure the extent of the Milky Way system, during the same seven year period less than 150 papers were published focusing on Cepheid variable stars. A nice illustration of the general interest in novae at the time can be found in the popular review of the nova phenomena by Francis (1919). Even the *Chicago Daily Tribune* was taking note of the latest developments by publishing a poem celebrating the discovery by Zygmunt Laskowski, an amateur astronomer, of one of the brightest Galactic novae on record—Nova Aquilae, which reached magnitude –0.5. The poem, penned by B. L. Taylor, and published on 1918 June 18 reads:

> Twinkle, twinkle, fickle star!
> Fading, fading fast you are.
> Yesterday so bright and fair,
> Pretty soon you won't be there.

It is likely that the intense interest in novae was due in large part to the mysterious nature of these objects. As the 1920s were getting underway, the physical mechanism responsible for the nova phenomena was completely unknown. The prevailing theory of nova eruptions at the time was that they were the result of the passage of stars through (gaseous) nebulae (Hagen 1921).

By early 1923 a total of 22 novae had been discovered in the Great Nebula in Andromeda. The most recent nova, Nova #22, had been found by Milton Humason (1891–1972)[11] on a plate taken on February 15 of that year (Humason 1923). The nova was located just under 100 arc seconds from the center of the Galaxy, closer than any other transient source since S And, the ultra-luminous "nova" of 1885.

Perhaps partially inspired by Humason's recent discovery, in the fall of 1923, famed astronomer Edwin Powell Hubble (1889–1953) initiated a long-term photographic survey for novae in the Andromeda nebula using the most powerful telescope in the world at the time, the 100 in Hooker telescope atop Mount Wilson. Hubble took his first plate on the night of 1923 October 4–5 under poor

[10] The comparison was determined by the number of papers with "nova," "novae," "nebula," or "nebulae" in the title.

[11] Humason, a well-known figure in early twentieth-century astronomy, was a former janitor and mule-team driver at the Mount Wilson Observatory who possessed no formal education. Nevertheless, his passion for astronomy and his skill as an observer led to his appointment (by Hale) to the scientific staff of Mount Wilson in 1919. He would go on to assist Hubble in observations leading to the discovery of the expansion of the Universe.

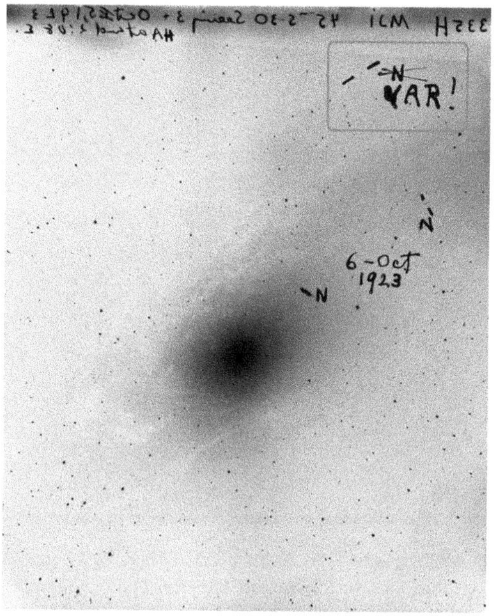

Figure 2.3. A reproduction of the historical Mount Wilson 100 inch plate H335H taken by Edwin Hubble on the night of 1923 October 5–6. The star in the upper right corner of the plate, initially thought to be a nova, turned out to be the first Cepheid variable star discovered in M31. (Image courtesy of the Carnegie Institution for Science.)

observing conditions. However, it was on the following night, Friday, October 5–6, when Hubble exposed a plate (H335H shown in Figure 2.3) that would soon resolve the long-standing confusion regarding the nature of the spiral nebulae. The plate, which is widely regarded as perhaps the most significant single observation in the history of cosmology, shows the discovery of the first Cepheid variable associated with a spiral nebula. Initially, Hubble had identified three objects on the plate that he believed to be novae. Upon closer inspection, and after comparing the plates with several others in the Mount Wilson collection, he realized that the nova candidate furthest from the center of M31 was in fact not a nova, but a variable star, which became known as M31 V1. After scouring through the Mount Wilson plate collection and identifying the star on numerous plates, he was able to establish that the variable star was indeed a Cepheid variable with a period of 31 days. The resulting light curve is shown in Figure 2.4. A simple comparison with Shapley's period–luminosity (P–L) relation for nearby Galactic Cepheids revealed that M31 V1 must lie at a distance of some 300,000 parsecs (1 pc = 3.26 light-years), and thus must be located well outside the confines of the Milky Way. Despite the Earth-shaking import of the discovery, Hubble characteristically did not rush to publish his results. Among other things, Hubble was worried that maybe the Cepheid was heavily obscured and thus its distance significantly overestimated. Instead, he took his time to carefully confirm his discovery through the discovery and analysis of several additional Cepheids in Andromeda.

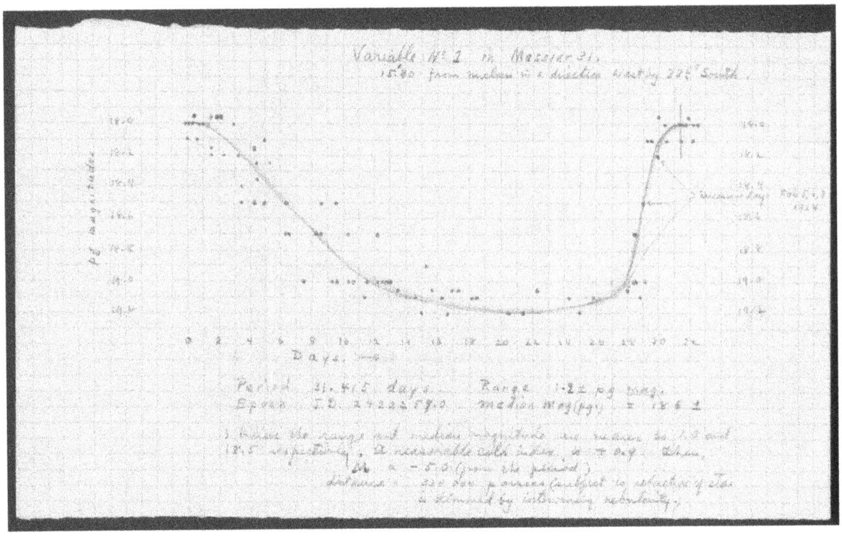

Figure 2.4. The light curve for M31 V1 that Hubble included in his letter to Harlow Shapley on 1924 February 19. (Courtesy Harvard College Observatory. Records of Director Harlow Shapley. Correspondence [incoming and outgoing]. Enclosure from Letter from Edwin Hubble, 1924 February 19. UAV 630.22 [Box 9, Folder 71]. Harvard University Archives.)

By February of 1924 Hubble had discovered several more novae in M31, along with an additional Cepheid variable star. At this point, Hubble penned a letter to Shapley opening with "You will be interested to hear that I have found a Cepheid variable in the Andromeda Nebula ⋯" With the letter Hubble included a copy of the light curve for M31 V1. Shapley immediately recognized the significance of the result. According to Cecilia Payne (later Cecilia Payne-Gaposkin—a graduate student of Shapley's at the time who was in his office when he opened the letter) Shapley is reputed to have remarked "This is the letter that destroyed my Universe."

Despite having identified two Cepheids in M31, Hubble still was not yet prepared to go public and publish his results. Nevertheless, word spread privately and by the fall of 1924 it had leaked to the press. On 1924 November 24 the *New York Times* published an article entitled "Finds Spiral Nebulae are Stellar Systems; Dr. Hubble Confirms View That They Are 'Island Universes' Similar to Our Own. WASHINGTON, November 22.—Confirmation of the view that the spiral nebulae, which appear in the heavens as whirling clouds, are in reality distant stellar systems, or "island universes," has been obtained by Dr. Edwin Hubble of the Carnegie Institution's Mount Wilson observatory, through investigations carried out with the observatory's powerful telescopes."

Henry Norris Russell encouraged Hubble to formally present his results at the 33rd meeting of the American Astronomical Society (held in conjunction with a meeting of the American Association for the Advancement of Science; AAAS) that was to be held in Washington, DC starting December 30 and concluding on New Year's Day, 1925. There was a $1000 (~$15,000 in today's dollars) prize from the AAAS available for the best paper presented at the meeting, and Russell felt confident that Hubble's work would be worthy of the prize. Hubble, however, remained reluctant to formally present his results even though it had been more than a year since the discovery of M31 V1. His reticence stemmed in large part from the fact that the extragalactic nature of spiral nebulae directly contradicted the proper motion measurements of Adriaan van Maanen, Hubble's senior colleague at Mount Wilson. If spiral nebulae were at the great distances implied by the Cepheid observations in the spirals (~1,000,000 light years in the case of M31 and M33), then the spiral nebulae would be too large to rotate as fast as implied by van Maanen's proper motion measurements.

Despite his initial reservations, Hubble did finally send a paper "Cepheids in Spiral Nebulae," which Russell read for Hubble at the meeting. Hubble's paper went on to share the Second AAAS prize (the Newcomb Cleveland prize) with a zoologist named Lemuel Roscoe Cleveland.[12] After the meeting the consensus among astronomers was that Hubble's work had convincingly shown that the spiral nebulae were indeed extragalactic. The long debate over the nature of the spiral nebulae was at last resolved.[13] They were "island universes," separate and apart from the Milky Way. The modern term "galaxy" eventually caught on, despite that fact that Hubble never adopted the term, and continued to refer to the spiral nebulae as "extragalactic nebulae" until his death in 1953.

As a side note, it is interesting to reflect that by the early 1920s observations of both Galactic and M31 novae had reached the point where one could reasonably have concluded that M31 must lie beyond the confines of the Milky Way system. As already discussed, Curtis (1917) had been the first to point out that the apparent magnitudes of typical novae in the field of the Andromeda nebula were approximately 10 mag fainter, and thus 100 times more distant that the typical Milky Way nova. Another important and influential player in the early study of novae was the Swedish astronomer, Knut Emil Lundmark (1889–1958). While still a student at Uppsala University, Lundmark determined a distance of 650,000 lt-yr to the Andromeda nebula by comparing the apparent magnitudes of the M31 novae known at the time with absolute magnitudes that he had determined for a sample of Milky Way novae with measured distances. After receiving his degree in 1920, Lundmark spent most of the next decade working at the Lick and Mount Wilson observatories. During this time he would continue to refine his estimates for the absolute magnitudes of Galactic novae. Lundmark (1922a, 1922b) used secular parallax to estimate a mean absolute magnitude of $m_{pg} = -7.1$ for a sample of 11 Galactic novae. In cases where individual parallaxes were available, he found a somewhat fainter mean absolute magnitude of $M_{pg} = -6.2$. On the other hand, typical apparent photographic magnitudes of novae in the Andromeda Nebula

[12] Matching surnames is a coincidence. Cleveland's contributions were entitled: "The ability of termites to live perhaps indefinitely on a diet of pure cellulose" and "The effects of starvation and oxygenation on the symbiosis between termites and their intestinal protozoa, together with the toxicity of oxygen for free-living and parasitic protozoa."

[13] An excellent and more detailed account of the exchanges prior to the 33rd AAS meeting, including the relevant correspondence between Joel Stebbins, then secretary of the AAS, Russell and Hubble, can be found in Berendzen & Hoskin (1971).

(S And, excluded) were in the range of 16th to 17th at maximum light. Thus, a comparison of the two values immediately suggested a distance modulus (m - M) for the Andromeda Nebula $\mu_{pg} \sim 23$—a value quite close to that derived a few years later by Hubble using Cepheid variables.

At the end of the day, credit for definitively establishing the extragalactic nature of the Andromeda Nebula (and by extension the other smaller and fainter spiral systems) goes to the Cepheids—both because they were thought to be better understood, and because of the lingering uncertainty surrounding the nature of objects such as "nova" Andromeda 1885, S And.

2.5 Hubble's Survey: The Mount Wilson Nova Collection

While continuing his work on Cepheids in spiral nebulae, Hubble continued to discover novae in M31. Between the fall of 1923 and the end of 1927, Hubble had added 63 nova candidates to the 22 that had been known prior to the launch of his fall 1923 campaign. This collection of novae represented what could be considered the first "survey" of novae in M31, and was published in 1929. In his seminal paper "A spiral nebula as a stellar system, Messier 31" (Hubble 1929), Hubble summarizes the properties of all 85 M31 nova candidates discovered with the 60 and 100 inch reflectors at Mt Wilson from 1909 to the end of 1927. The paper also presents an analysis of the 40 Cepheid variables discovered since the fall of 1923. Figure 2.5

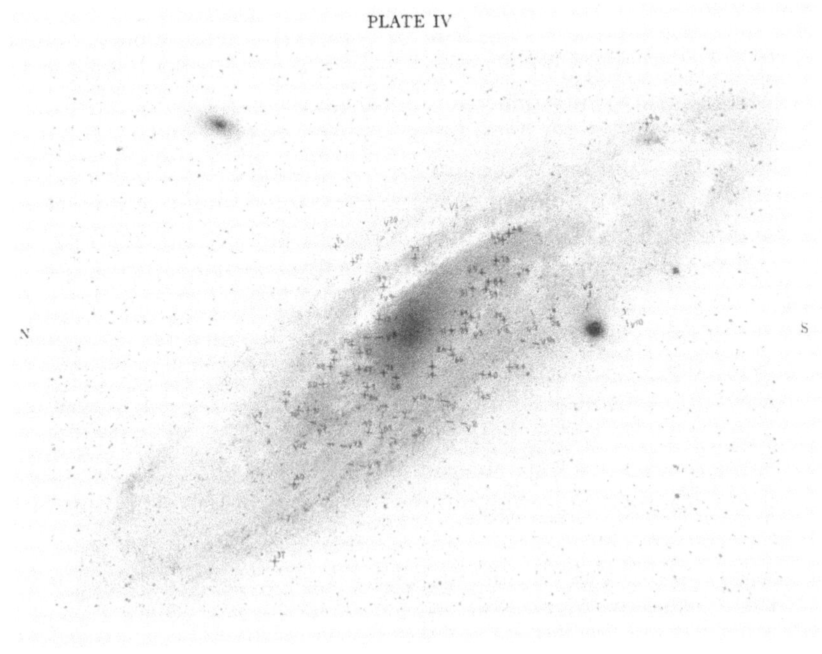

Figure 2.5. The positions of the 85 M31 novae from the Mount Wilson survey. Note the apparent drop-off in the number of novae near the Galaxy's nucleus. (Reproduced from Hubble 1929. © 1929. The American Astronomical Society. All rights reserved.)

shows the positions of the 85 novae from Hubble's survey superimposed on a photographic image of M31.

A number of fundamental insights concerning the nova population in M31 were realized as a result of Hubble's survey. The wealth of data enabled Hubble to study the light curves of many M31 novae. From these data Hubble was able to show that the general character of the light curves were quite similar to Galactic novae. Figure 2.6 shows the light curves of six novae that were observed near maximum light, while Figure 2.7 shows the frequency distribution of magnitudes at maximum. Hubble used these data in conjunction with his distance from the Cepheid P–L relation to derive an absolute magnitude at maximum light of −5.7 for M31 novae. This number was in good agreement with the estimates for many Galactic novae at the time (e.g., $\langle M \rangle = -6.2$; Lundmark 1922b). The apparent agreement was later shown to be a coincidence resulting from two unrelated misunderstandings. The mean absolute magnitudes for Galactic novae were systematically overestimated (luminosity underestimated) in part because of the failure to recognize the effect of interstellar extinction. The absolute magnitudes of M31 novae were also overestimated, but in this case due to the fact that the distance to M31 was underestimated (by approximately a factor of two) due to the failure to appreciate the luminosity difference between Population I and II Cepheids. Specifically, the P–L relation used by Hubble to determine the distance to M31 was calibrated using Pop II Cepheids in the Milky Way, which are less luminous that their Pop I counterparts observed by Hubble in the disk of M31.

The extended spatial coverage of the Mount Wilson plates allowed Hubble to make a study of the spatial distribution of novae in M31. Hubble concluded that

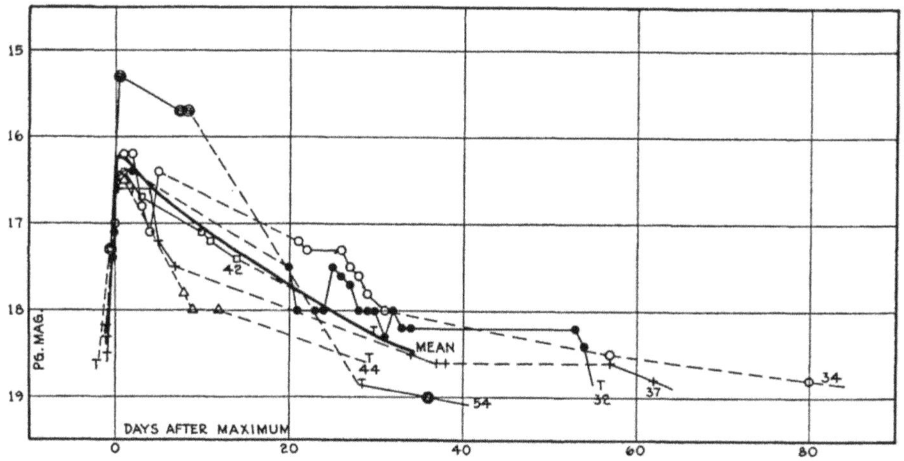

Figure 2.6. Light curves of six novae in M31. The average nova faded by ~0.05 mag per day, similar to Galactic novae. (Reproduced from Hubble 1929. © 1929. The American Astronomical Society. All rights reserved.)

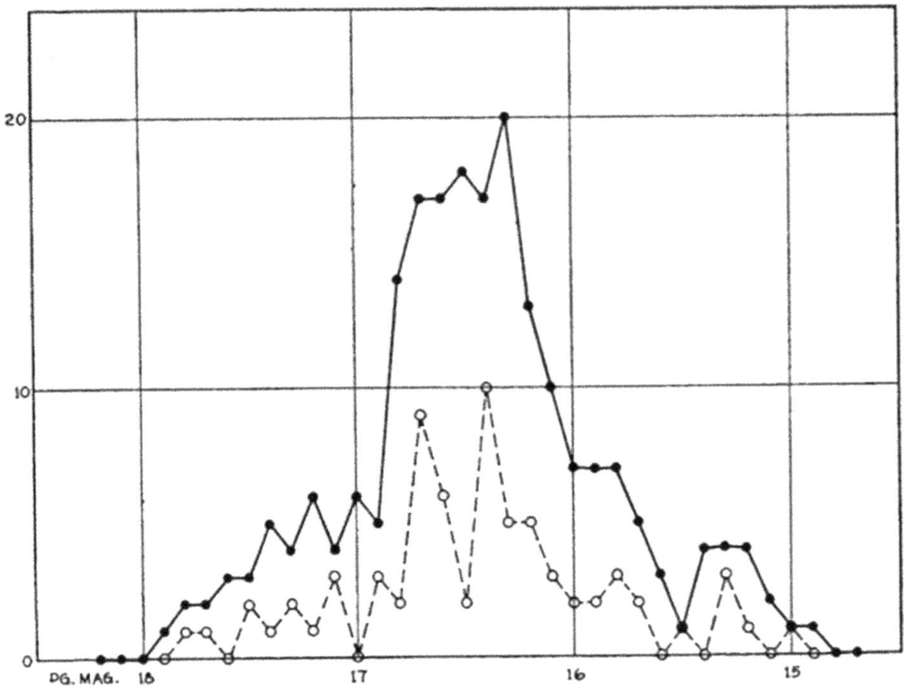

Figure 2.7. The frequency distribution of magnitudes at maximum for 71 novae in the Mount Wilson survey on M31. The filled points represent 0.3 mag wide bins, while the open circles represent 0.1 mag bins. The distribution peaks around $\langle m \rangle = 16.5$. (Reproduced from Hubble 1929. © 1929. The American Astronomical Society. All rights reserved.)

novae heavily concentrated in the bulge region, tended to avoid the dark regions between the spiral arms, and generally were distributed following the background light of the nebula. Hubble also noted a precipitous drop-off in the density of novae within ~4′ of the nucleus that he thought couldn't be explained by the bright background light alone. As we shall see this apparent nova "hole" would also be noted by Arp in his M31 survey two decades later. More recent surveys done in Hα light, however, such as that undertaken by Ciardullo et al. (1987) establish that the nova density follows the light all the way to the center of the Galaxy.

The wealth of data collected at Mount Wilson over the years between 1909 and 1927 also enabled Hubble to make a reasonably accurate estimate of the overall nova rate in the Andromeda nebula. During the last five years of the survey (1923–1927), the average observed nova rate was found to be approximately 16–17 per year. By assuming that the average nova would be detected on plates for a period of 15 days, and correcting for incompleteness due to poor seeing, spatial coverage, etc, Hubble estimated that "thirty per year is a reasonable estimate of the frequency of novae in M31." This number would be confirmed by subsequent studies by Arp (1956) and Rosino et al. (1989), and it remained the accepted rate throughout the twentieth century. In recent years estimates of the rate have crept up as it has become increasingly clear that previous studies likely underestimated the completeness of the earlier surveys. Modern studies put the actual rate at 2–3 times Hubble's value (see Chapter 6).

In conclusion, Hubble's early nova survey revealed several key insights into the nature of classical novae in M31: the character of the lightcurves were indistinguishable from novae found in the Milky Way, the nova density generally followed the background light of the Galaxy, and the nova rate was of order 30 per year, which was not inconsistent with the rate suspected at that time in the Milky Way. These conclusions, with the exception of the overall rate, which has been revised upward by a factor of approximately two, have all stood the test of time.

2.6 Novae and Supernovae

As we have seen, the failure to appreciate the distinction between novae and supernovae added considerable confusion in early attempts to establish the nature of the spiral nebulae. By the early 1920s, the apparent conflict between the brightnesses of the first "novae" to be discovered in spiral nebulae (e.g., S And in M31 and Z Cen in NGC 5253) and those of lower luminosity subsequently discovered in spirals by Ritchey (1917a, 1917b) was widely recognized. Lundmark is credited as the first to suggest (in 1920) a class of "Giant Novae," followed by Curtis (1921) who remarked in the written record of the *Great Debate*: "It seems certain, for instance, that the dispersion of the novae in the spirals, and probably also in our galaxy, may reach at least ten absolute magnitudes, as is evidenced by a comparison of S And with faint novae found recently in this spiral. A division into two magnitude classes is not impossible."

Evidence for two classes of novae became even more compelling during the decade that followed with Hubble's (1929) compilation of 86 novae in M31. Of the 86 novae, 85 were "common" novae, comparable to those seen in the Galaxy, with the one exception being the luminous nova seen in 1885, S And. By the time that Hubble had completed his compilation of novae in M31, a total of 19 "novae" had been spotted in other, more distant galaxies (see Table 2.1 and Figures 2.8 and 2.9). In his classic 1929 paper, Hubble remarked "The occasional novae which have appeared in the smaller extra-galactic nebulae and which have attained luminosities that are respectable fractions of the total luminosities of the nebulae themselves are clearly a different sort of object from the normal novae in M 31 and 33. They are generally classed with S And, the nova of 1885 in M 31, as a rare and peculiar type."

Given that the host galaxies of these 19 novae were more distant than M31, the observed apparent magnitudes suggested that these novae were considerably more luminous than the common novae typically found in M31 and the Galaxy. Indeed, the apparent magnitudes of these novae rivaled that of the entire nebula in which they arose. By the early 1930s work by Hubble and others had established that the absolute magnitudes of the extragalactic nebulae were clustered near $M(\mathrm{vis}) \sim -14.9$ (Hubble & Humason 1931), and therefore that the extragalactic novae must have comparable luminosities.

By the early 1930s the division of novae into two separate luminosity classes had finally become widely accepted. A new term "super-novae" was coined by Fritz Zwicky and Walter Baade circa 1931 in seminars given at the California Institute of

Table 2.1. Extragalactic Super-novae Prior to 1930

SN	Nebula	Disc. Date	Mag	Discoverer	References
1885A	NGC 0224	1885 Aug 17	5.8	Hartwig	...
1895A	NGC 4424	1895 Mar 16	12.5	Wolf	...
1895B	NGC 5253	1895 Jul 7	8.0	Fleming	...
1901A	NGC 2535	1901 Jan 10	14.7	Reinmuth	...
1901B	NGC 4321	1901 Mar 17	13.5	Curtis	Curtis (1917)
1907A	NGC 4674	1907 May 9	13.5	Luyten	...
1909A	NGC 5457	1909 Jan 16	13.5	Wolf	...
1912A	NGC 2841	1912 Feb 19	...	Pease	Pease (1917a, 1917b)
1914A	NGC 4321	1914 Mar 2	14	Curtis	Curtis (1917)
1915A	NGC 4527	1915 Mar 20	15	Curtis	Curtis (1917)
1917A	NGC 6946	1917 Jul 19	14.6	Ritchey	Ritchey (1917b)
1919A	NGC 4486	1919 Feb 24	12	Balanowsky	...
1920A	NGC 2608	1920 Jan 1	13.2	Wolf	...
1921A	NGC 4038	1921 Mar 7	...	Hubble, Duncan	...
1921B	NGC 3184	1921 Apr 6	13.5	Zwicky	...
1921C	NGC 3184	1921 Dec 5	11.0	Jones	...
1923A	NGC 5236	1923 May 9	14.0	Lampland	...
1926A	NGC 4303	1926 May 9	14.0:	Wolf, Reinmuth	...
1926B	NGC 6181	1926 Jun 17	14.8	Van Maanen	Van Maanen (1926)

Technology (Caltech). Baade and Zwicky would go on to make many important contributions to astronomy during their storied careers, and although they were both recognized as brilliant scientists, their personalities couldn't have been more different.

Wilhelm Heinrich Walter Baade (1893–1960) was born in Germany. After receiving his doctorate at Göttingen in 1919, Baade spent most of the next decade at the Hamburg Observatory. Despite having a relatively secure position there, Baade had dreamed since early in his career of conducting research with the world's biggest telescopes at the Mount Wilson Observatory. However, it wasn't until 1927 when he made his first trip to America, that he was finally able to visit Mount Wilson. While there Baade's congenial manner made a favorable impression with the Mount Wilson staff, including the director Walter S. Adams. Baade's research interests were broad, and he developed an early interest in novae, and their potential use as distance indicators to the spiral nebulae. In 1927 he introduced the term "Hauptnova" (chief nova) to refer to the very luminous novae such as S And. Four years later, in 1931, Baade's dream was realized when he was offered a permanent position on the Mount Wilson staff.

In many ways the polar opposite of Baade, Fritz Zwicky (1898–1974), was anything but soft-spoken and congenial. Zwicky was born in Bulgaria in 1898, but raised in Switzerland, where he received his education at the Swiss Federal Polytechnic, known today as ETH Zurich. After completing his doctorate in 1922, Zwicky immigrated to the US in 1925 and took a position at Caltech working

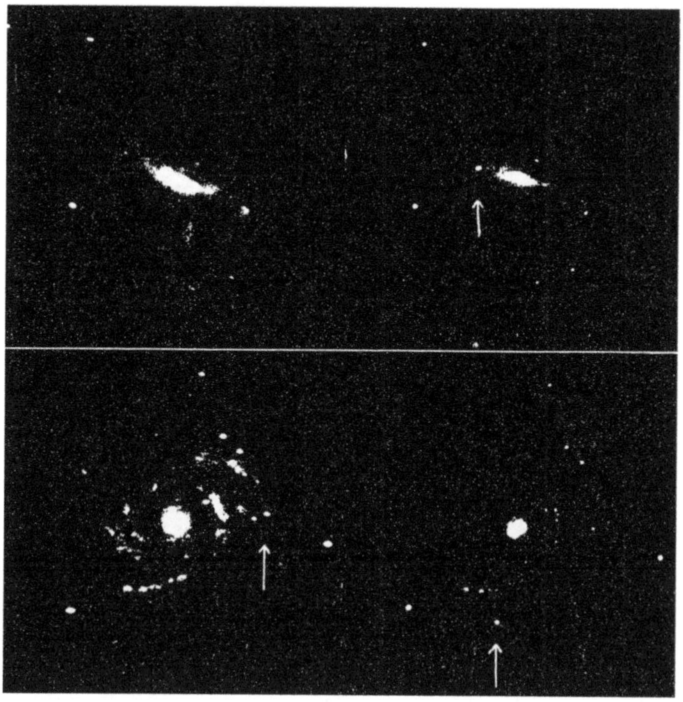

Figure 2.8. Three early "novae" discovered in spiral nebulae. (Top) An apparent nova discovered in NGC 4527. The left photograph was taken on 1901 May 8, while the right photograph shows the "nova" detected on 1915 April 16. (Bottom) Two apparent novae in NGC 4321. The photograph on the left shows an apparent nova from 1901 April 19, while the photograph on the right shows another "nova" in the same galaxy from 1914 March 2. All three apparent novae were later recognized as supernovae (SN 1915A, SN 1901A, SN 1914A). (From Curtis 1917. Reproduced with permission of the University of California Press.)

with Nobel laureate Robert Millikan (1868–1953). While at Caltech, Zwicky not only developed a reputation for being a brilliant scientist, but also for being a rather surly and abrasive individual who did not get along well with many of his colleagues. Although Zwicky began his career as a theoretical physicist working on the quantum mechanical structure of crystals, in the mid 1920s his interests turned to astrophysics; in particular, the study of cosmic rays with Millikan, and eventually to novae.

Their mutual interest in the most luminous novae brought Baade and Zwicky together shortly after the former astronomer's arrival at Mount Wilson. However, it wouldn't be until January of 1934 when the term "super-nova" was first used by Baade and Zwicky in a pair of remarkable papers: "On Super-novae" and "Cosmic Rays from Super-novae," published back-to-back in the Proceedings of the National Academy of Sciences (Baade & Zwicky 1934a, 1934b). In the first paper, Baade and Zwicky opened by stating: "The extensive investigations of extragalactic systems during recent years have brought to light the remarkable fact that there exist two well-defined types of new stars or novae which might be distinguished as common novae and super-novae. No intermediate objects have so far been observed." Baade and Zwicky went on to estimate that $\sim 10^{51}$ ergs were liberated during a super-nova,

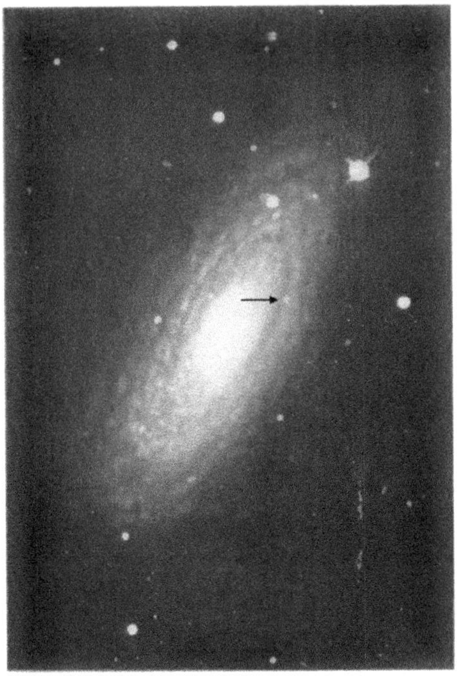

Figure 2.9. An apparent nova in NGC 2841 discovered on 1912 February 19 by Pease (1917a, 1917b). As with the three "novae" shown in Figure 2.8, the object would later be recognized as a supernova (SN 1912A). (Reproduced from Pease 1917b. © 1917. The American Astronomical Society. All rights reserved.)

and that a considerable fraction of the star's initial mass must be converted into energy to power such an explosion.

In their subsequent paper, Baade & Zwicky (1934b) proposed that the acceleration of charged particles and the production of hard radiation in super-novae explosions were the source of cosmic rays. In the same paper, they were the first to propose a qualitative mechanism for the production of a super-nova. Specifically, they wrote: "With all reserve we advance the view that a super-nova represents the transition of an ordinary star into a *neutron star*, consisting mainly of neutrons. Such a star may possess a very small radius and an extremely high density." The ideas put forward by Baade & Zwicky, although speculative at the time, have been proven remarkably prescient.

On the heels of the Baade and Zwicky papers, Lundmark (1935) summarized the observational properties of novae as they were understood at the time. Rather than two classes, Lundmark divided novae into *three* principal classes, A: The "upper-class novae" or "super-novae" ($M_{\rm max} \sim -15$), B: "middle-class," or ordinary novae ($M_{\rm max} \sim -7$), *and* C: "lower-class," or dwarf novae ($M_{\rm max} \sim +3$ or 4). The existence of class C, the dwarf novae, was deemed to be "not well established" at the time. He remarked that there were "only sporadic and uncertain observational data indicating the existence of this subdivision."

By far the most novae observed in M31 and the Galaxy fell into class B, with S And being the sole exception in M31. The "dwarf novae" that Lundmark was referring to are not the same objects (e.g., U Gem, SS Cyg, etc) that we currently classify as dwarf novae (see Chapter 4). He went on to make a remarkably accurate prediction concerning the rate of novae and supernovae in the Galaxy: "It seems that the number of outbursts of ordinary novae in our galaxy is at least 50 a year, but probably much higher.[14] The number of upper-class Novae can not be very accurately estimated but it should be one in 50 years or so. Thus, the ordinary novae are at least 2500 times more frequent than the super-Novae."

By the end of the decade, the hyphen had been dropped from the name, and the modern term "supernova" had become generally accepted, replacing the older ones—"giant novae," "exceptional novae," and "Haupt novae," that had been introduced previously by Lundmark, Hubble, and Baade, respectively. The prefix "super" was apparently quite popular in the 1920s and 1930s, with the introduction of terms such as "superpower" and "supertanker" appearing in 1921 and "supermarket," "superhighway," and "supersonic," appearing in 1925 (Koenig 2005). It wouldn't be the first nor the last time that the popular culture influenced astronomical nomenclature.

2.7 The Maximum-magnitude Rate-of-decline Relation

As we have seen, novae played a prominent role in early attempts to determine whether the spiral nebulae were extragalactic. In particular, a qualitative comparison between the observed properties (mainly peak apparent magnitude) of Galactic and spiral novae, strongly suggested that the spirals were outside the Milky Way. However, little was known about the detailed nova luminosity function. In particular, the initial confusion between novae and supernovae, as exemplified by the extreme luminosity of S And compared with other novae in M31, confounded any attempt to use novae as quantitative distance indicators.

Early attempts to use novae as distance indicators implicitly assumed that the intrinsic dispersion in the luminosities of "common novae," (Lundmark's class B) was small. By 1935 more than one hundred novae had been identified in M31. From these data Lundmark (1935) determined a dispersion in absolute magnitude at maximum light of just $\sigma = 0.53$ mag for the ordinary novae, of which he estimated that $\sigma = 0.42$ mag was not due to observational uncertainty, but was intrinsic to the novae. In reflecting on the early observations, Lundmark remarked that "The empirical fact that $M_{\rm max}$ is rather constant for Novae has given them an important application as distance indicators. The scanty data at hand around 1920 pointed in the direction that $\sigma_{\rm max}$ should at least be smaller than 1 [mag] but probably considerably smaller. Already at this time the right order of magnitude for the distance to the Andromeda galaxy could be determined."

The first major steps toward the use of novae as a quantitative distance indicator came in the mid 1930s when Fritz Zwicky published a pair of papers that attempted to correlate nova luminosities with observed properties of the eruption. The first paper (Zwicky 1936a) explored the theoretical basis for a correlation between a

[14] This is in close agreement with the latest estimate for the Galactic nova rate (Shafter 2017).

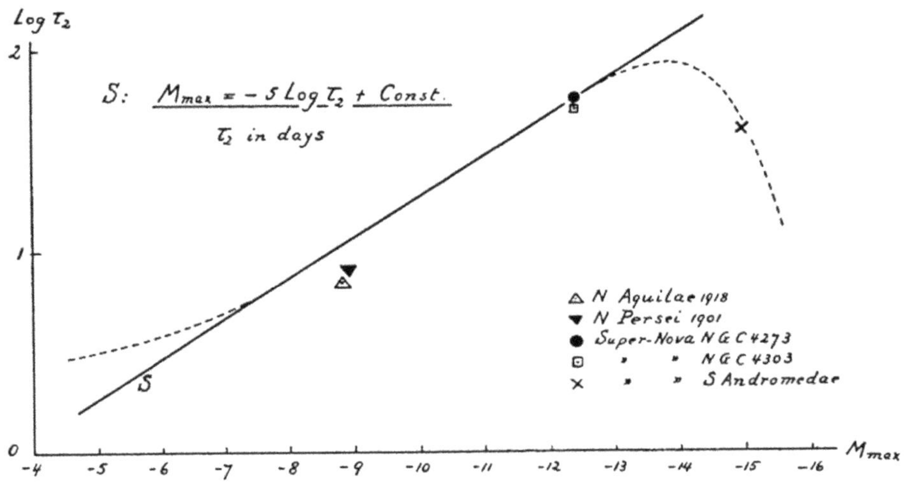

Figure 2.10. Zwicky's life–luminosity relation for novae. Failure to differentiate between novae and super-novae resulted in brighter "novae" having longer lifetimes. This is the opposite behavior to the MMRD relation for novae. (From Zwicky 1936b. © 1936. The Astronomical Society of the Pacific. All rights reserved. Printed in U.S.A.)

nova's absolute magnitude and its expansion velocity that had been first suggested by Adams (1920). In the second paper, Zwicky (1936b) explored the relationship between a nova's absolute magnitude and its "lifetime," $t_{\Delta M}$, where ΔM was taken as 2 mag. This was the first time that the so-called t_2 time was employed to describe the fade rate of novae. Figure 2.10 shows Zwicky's initial attempt to define a relation between the maximum luminosity and decline rate for novae. The first thing to note is that despite his understanding that there were two kinds of novae, ordinary novae and super-novae, Zwicky did not distinguish between the two when formulating his "life–luminosity" relation for novae. As a result, he found that brighter novae (the super-novae) faded more slowly than the fainter "ordinary" novae.

The first suggestion of a possible relation between the peak luminosity of a common nova and its rate of decline came in a paper by Dean B. McLaughlin (1901–1965) entitled: The Luminosities of Novae (McLaughlin 1942), where he stated "Since Nova Aquilae and Nova Persei, whose luminosities were of the highest, both were fast novae, there is some suggestion of a connection between luminosity and rate of decline, or a 'life-luminosity relation.' But there is no reason to believe that it is more than a statistical trend; a precise correlation is not implied." Unlike Zwicky, McLaughlin clearly distinguished novae from supernovae: "… But if there is a life–luminosity relation among common novae, it certainly does not extend to the supernovae. The latter have lifetimes which are *longer*,—not shorter,—than those of the fastest common ones."

Over the next few years McLaughlin sought to strengthen his argument for a "life–luminosity" relation demonstrating that the intrinsically brighter novae faded faster than dimmer novae. Measuring nova fade rates was relatively straightforward, but the determination of reliable absolute magnitudes was complicated by the

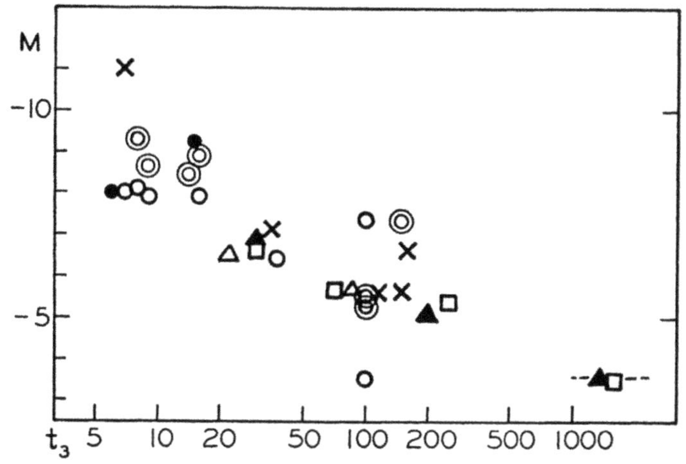

Figure 2.11. McLaughlin's maximum-magnitude versus rate-of-decline (MMRD) relation for novae. The abscissa is the number of days to decline by 3 mag, t_3, while the ordinate is the absolute magnitude, M. Data symbols: double circles, nebular expansion; open circles, distance from interstellar calcium lines; solid circles, T Coronae and T Scorpii; crosses, distance from galactic rotation effect; solid triangles, Andromeda galaxy; open triangles, LMC; squares, Sagittarius. (From McLaughlin 1945. © 1945. The Astronomical Society of the Pacific. All rights reserved. Printed in U.S.A.)

uncertain effects of interstellar extinction. Undaunted, McLaughlin compiled the most reliable data on the absolute magnitudes and fade rates of a number of Galactic, Large Magellanic Cloud (LMC) and M31 novae. This work culminated in a seminal paper where he laid out a compelling case for a life–luminosity relation for common novae (McLaughlin 1945). This relation, which would become known as the maximum-magnitude vs rate-of-decline relation (hereafter the MMRD relation), is shown in Figure 2.11.

It is an interesting historical note that in his 1945 paper, McLaughlin recognized that the relationship broke down for the recurrent novae known at the time. In McLaughlin's words "The status of the recurrent novae relative to the life–luminosity relation is very uncertain, ⋯" This statement would turn out to be prescient, as the latest research calling into question the existence of the MMRD (see Chapter 8), recognizes that recurrent novae seem to occupy the "faint and fast" quadrant in the MMRD plane.

References

Adams, W. S. 1920, ApJ, 51, 126
Adams, W. S. 1939, BM, 21, 181
Arp, H. C. 1956, AJ, 61, 15
Baade, W., & Zwicky, F. 1934a, PNAS, 20, 254
Baade, W., & Zwicky, F. 1934b, PNAS, 20, 259
Berebdzen, R., & Hoskin, M. A. 1971, ASPL, 10, 425
Brèmond, A. G. 2009, JAHH, 12, 72
Ciardullo, R., Ford, H. C., Neill, J. D., et al. 1987, ApJ, 318, 520

Clerke, A. M. 1905, The System of the Stars (2nd ed.; London: Adam Charles Black), 349
Curtis, H. D. 1915, PASP, 27, 214
Curtis, H. D. 1917, LicOB, 9, 108
Curtis, H. D. 1921, BuNRC, 2, 171
Fleming, W. P. S., & Pickering, E. C. 1912, AnHar, 56, 165
Francis, V. 1919, PA, 27, 7
Gingrich, O. 1988, JHA, 19, 201
Hagen, J. G. 1921, ApJ, 54, 229
Hoskin, M. A. 1976, JHA, 7, 169
Hubble, E. 1929, ApJ, 69, 103
Hubble, E., & Humason, M. 1931, ApJ, 74, 43
Humason, M. 1923, PASP, 35,
Koenig, T. 2005, in ASP Conf. Ser. 342, Supernovae as Cosmological Lighthouses, ed. M. Turatto, S. Benetti, L. Zampieri, & W. Shea (San Franciso, CA: ASP), 53
Lundmark, K. 1922a, PAAS, 4, 368
Lundmark, K. 1922b, PASP, 34, 207
Lundmark, K. 1935, MeLuS, 74, 1
McLaughlin, D. B. 1942, PA, 50, 233
McLaughlin, D. B. 1945, PASP, 57, 69
Pease, F. G. 1917a, PASP, 29, 213
Pease, F. G. 1917b, ApJ, 46, 24
Ritchey, G. W. 1917a, BHarO, 641, 1
Ritchey, G. W. 1917b, PASP, 29, 210
Ritchey, G. W. 1917c, PASP, 29, 257
Rosino, L., Capaccioli, M., D'Onofrio, M., & della Valle, M. 1989, AJ, 97, 83
Shafter, A. W. 2017, ApJ, 834, 196
Shapley, H. 1917, PASP, 29, 213
Shears, J. 2011, JBAA, 121, 36
Slipher, V. A. 1914, PA, 22, 19
Thompson, L. A. 2011, arXiv: 1108.4864v2
Van Maanen, A. 1926, PASP, 53, 125
Van Maanen, A. 1916, ApJ, 44, 210
Zwicky, F. 1936a, PNAS, 22, 457
Zwicky, F. 1936b, PASP, 48, 191

Chapter 3

Early Photographic Surveys

3.1 Zwicky's Pioneering Supernova Search

In the early twentieth century Bernhard Woldemar Schmidt (1879–1935) had become a well-known and successful optician in Germany. However, like many of his countrymen, he fell on hard times after World War I and the subsequent economic collapse of Germany in the post-war years. After unsuccessful attempts to resurrect his old business, or start a new one, Schmidt eventually took a position at the Hamburg Observatory in 1927 where he became acquainted with Walter Baade. During his time at Hamburg, Schmidt developed what would turn out to be his most important invention, a wide-field camera capable of photographing large areas of the sky. Upon his arrival to Mount Wilson in 1931, Baade brought news of the revolutionary new instrument, the eponymous Schmidt Telescope, with him.

After the spectacular nature of supernovae was recognized in 1935, Fritz Zwicky became interested in conducting a large-scale search for supernovae in extragalactic nebulae. With the assistance of Baade, Zwicky was able to secure sufficient funding from Hale, then director of Caltech's newly-established Palomar Observatory, to build an 18 inch Schmidt telescope and place it atop Mount Palomar (see Figure 3.1). The Schmidt telescope became operational in 1936 September, and for the next 13 years would be the only operational research telescope at the Observatory until the crown jewel of Palomar, the famed 200 inch reflector, would see first light in 1949.

Starting on 1936 September 6, Zwicky began his supernova survey that included the fields containing several nearby galaxies (M33, M31, M81, etc), as well as areas containing large numbers of more distant galaxies, such as those in the Virgo cluster. At the end of a little more than a year, Zwicky had discovered a total of three supernovae within approximately 3000 galaxies that he had monitored. When

doi:10.1088/2514-3433/ab2c63ch3

Figure 3.1. Fritz Zwicky using the Palomar 18 inch Schmidt Telescope circa 1936. (Courtesy of the Caltech Archives.)

corrected for the temporal coverage (what he called the "control value"), Zwicky (1938) estimated a supernova frequency of roughly one supernova per galaxy per 612 years.[1] By 1940 January Zwicky had discovered an additional nine supernovae and revised his rate to one supernova per galaxy per 430 years (and one supernova per galaxy per 359 years when only the brighter Shapley-Ames; Shapley & Ames 1932; galaxies were included).

During the late 1930s the irascible Zwicky and the genteel Baade drifted apart. The last paper the two were to publish together was a paper on the light curves of two supernovae discovered during the first year of the Palomar supernova survey (Baade & Zwicky 1938). By the early 1940s it was clear that the two had experienced a major falling out. According to Jesse Greenstein, then head of the newly-established Astronomy program at Caltech: "He [Zwicky] and Baade split violently. They were a dangerous pair to put in the same room; in fact. Zwicky called Baade a Nazi, which he wasn't. And Baade said he was afraid that Zwicky would kill him."[2] During the war years, Baade would go on to establish the existence of two stellar populations through his work on M31. Zwicky would continue his supernova searches for most of the rest of his life, and by the time he died in 1974, he had discovered a total of 122 supernovae by himself (and one more, SN 1963J, with Paul Wild), a record that stood until 2009 when it was eclipsed by British amateur astronomer Tom Boyles.

[1] This value would later be revised to one per galaxy per 457 years when it was realized that one questionable variable source omitted from the original analysis was indeed a supernova.

[2] Greenstein, Jesse L. Interview by Rachel Prud'homme. Pasadena, California, 1982 February 25–March 23. Oral History Project, California Institute of Technology Archives. Retrieved 2018 November 16 from: http://resolver.caltech.edu/CaltechOH:OH_Greenstein_J.

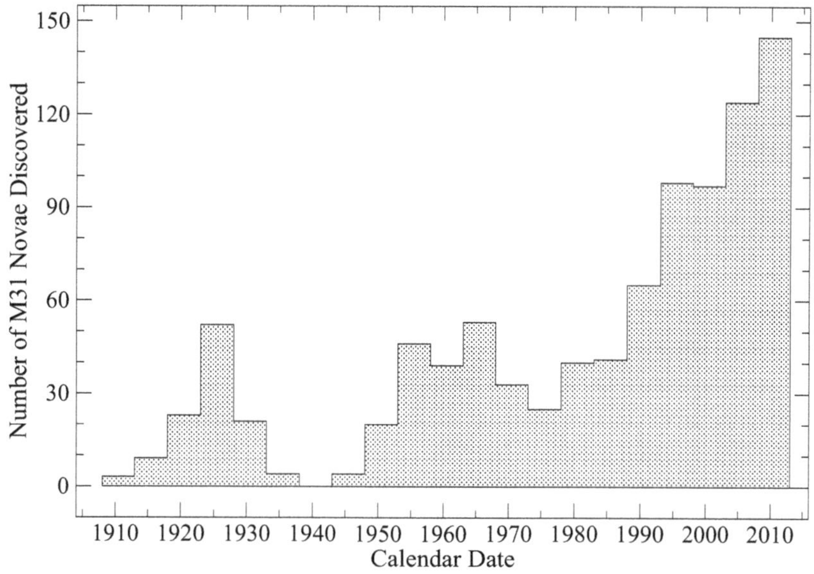

Figure 3.2. History of nova discoveries in M31. Discoveries dropped significantly during World War II, and then accelerated after CCD detectors became commonplace in the 1980s.

3.2 Arp's M31 Nova Survey

Despite Zwicky's supernova program, and McLaughlin's pioneering work in developing the MMRD relation in the 1940s, interest in discovering and studying new "common" novae waned considerably throughout most of the 1930s, and the number of novae discovered in M31 declined precipitously. The dearth of discoveries persisted throughout the following decade and into the early 1950s, as can be seen in Figure 3.2.

Interest in novae rebounded in the early 1950s when the first dedicated survey for novae in M31 was undertaken by Halton C. (Chip) Arp (1927–2013). Working under the direction of Baade, Arp had just received his doctorate at Caltech in 1953. With encouragement from Hubble, Baade convinced Arp that a dedicated survey for novae in M31 was long overdue. Despite not being thrilled with the project.[3] Arp spent much of the next two years atop Mount Wilson observing M31 with the Observatory's 60 inch telescope.

A principal goal of the survey was to study the nova light curves in detail, which required frequent monitoring. Hubble was able to secure the Mount Wilson 60 inch telescope for two observing seasons, where a total of $\sim$1000 5 $\times$ 7 inch plates were obtained on 290 usable nights between 1953 June and 1955 January. By the time the

[3] In an excerpt from his oral history interview for the American Institute of Physics, Arp recalled his feelings at the time on the project: "... Hubble and Baade, one of the few things they agreed on, they wanted a nova search done, particularly Hubble, in M31. That meant going up on Mt. Wilson to live for two years and observe almost every night, just picking up novae, I think I discovered thirty in two years or something like that. I wasn't terribly excited about it."

survey wrapped up, a total of 30 novae had been discovered (Arp 1956). The nearly continuous coverage produced many remarkably complete light curves, where maximum light was missed by more than a day for just five novae. Durations of nova visibility (defined as the number of days the nova remained brighter than 20th magnitude) and estimates of nova fade rates (based on the average decline rate over the first two magnitudes from maximum) were measured, and demonstrated that the novae in his M31 sample were quantitatively similar to those of Galactic novae. Regarding the M31 novae, Arp (1956) concluded that "… their absolute magnitudes are in exact accordance with luminosities of well-determined [G]alactic novae. Light curves and outburst energies are also strictly comparable."

Figure 3.3(a) shows the MMRD relation for the complete sample of Arp's M31 novae. In good agreement with their Galactic counterparts (McLaughlin 1945), the novae displayed a clear correlation with the brightest novae fading considerably faster than the fainter ones. The one clear outlier was Nova #4, which was considerably fainter at maximum light compared with other novae with similar fade rates. Arp explained the discrepancy by noting the color index for Nova #4 was considerably "redder" than usual near maximum light[4] suggesting that this nova was heavily obscured by dust.

In addition to providing a large sample of complete nova light curves, the nearly continuous coverage of Arp's survey allowed him to explore the distribution of nova peak luminosities. Figure 3.3(b) shows the distribution of apparent magnitudes at maximum light. The distribution is bimodal with a bright peak centered at a photographic magnitude $m_{pg} \sim 16.0$ and the fainter peak near $m_{pg} \sim 17.5$, suggesting the possibility that there may be two distinct populations of novae in M31. The small number of novae in the sample, however, made this interpretation far from

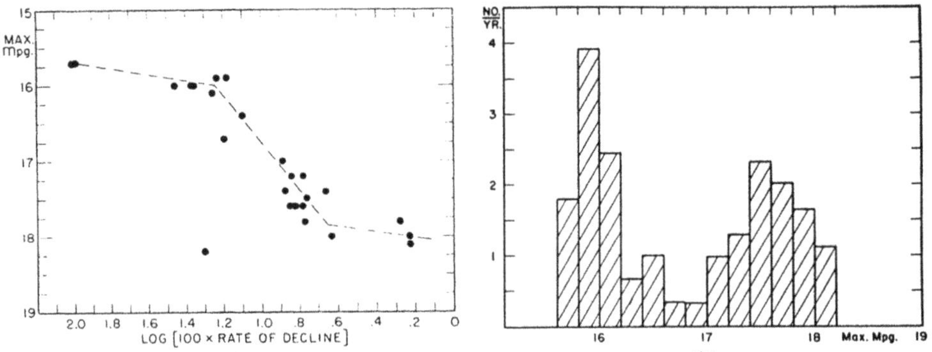

(a) (b)

Figure 3.3. (a) The MMRD relation from Arp's M31 nova survey. (b) The apparently bimodal distribution of magnitudes at maximum light, which has been used to support the notion that there are two populations of novae. (Reproduced from Arp 1956. © 1956. The American Astronomical Society. All rights reserved.)

[4] $B-V \sim 0.7$ versus $B-V \sim 0.2$.

certain. Indeed, later surveys, based on a significantly larger sample of novae would fail to reproduce a bimodal peak magnitude distribution.

By considering the combined sample from Hubble's survey and that of his own, Arp was also able to make a careful study of the spatial distribution of 112 novae in M31. The principal conclusions were similar to those drawn by Hubble two decades earlier. The novae exhibited a spatial distribution that was consistent with a population type "intermediate between a highly flattened disk population and a nearly spherical globular population." Like Hubble, Arp also concluded that the nova density dropped off within the inner 4′ of the Galaxy. Arp estimated that his images would be sensitive to novae to within 1.5′ of the nucleus, and thus he concluded that the decline in nova density was a real feature. As we shall see in Chapter 6, Hα images obtained with modern CCD detectors would later show that the "hole" in the nova distribution was in fact spurious.

3.3 Rosino's Survey at Asiago

The last of the major photographic nova surveys of M31 was undertaken by the Italian astronomer Leonida Rosino (1915–1997) in a long-term study that was to span more than 30 years. Rosino's observations began about the same time as Arp's survey, first using the 0.6 m telescope at Loiano (no novae were detected), and then in 1955 with the 1.22 m reflector at Asiago. Between 1955 September and 1963 September a total of 46 novae were identified on 9 × 12 cm blue-sensitive plates (Grubissich & Rosino 1955, 1957, 1959; Rosino 1964). In most cases the observational coverage was sufficient to characterize the light curve morphology. In agreement with the Hubble and Arp surveys, Rosino concluded that the M31 novae were qualitatively similar to Galactic novae. His MMRD relation, comparing the behavior of his M31 nova sample with Arp's sample and known Galactic novae, is shown in Figure 3.4. As a historical aside, in 1964 Rosino (1964) pointed out that the growing database of M31 novae made it possible to identify recurrent novae in that galaxy. He identified one possible such system, Nova #62 in Hubble's survey appeared to be quite close to the position of Nova #39 in his Asiago survey. Rosino, not having easy access to Hubble's original plates, was unable to confirm his suspicion and never pursued this possibility further, but did mention that the pair of novae "should be kept in future under control." Over 50 years later, Shafter and collaborators (Shafter et al. 2015), in their study of recurrent novae in M31, were finally able to confirm beyond a reasonable doubt that Rosino's nova #39 was indeed a recurrence of nova #62 discovered by Hubble on 1926 June 4.

The second phase of Rosino's M31 survey, covering 44 novae discovered from 1963 to 1970, added to the statistical sample of M31 novae, but provided few new insights concerning the M31 nova population (Rosino 1973). That said, the survey was noteworthy in that it identified the first recurrent nova to be confirmed outside the Milky Way. The 79th nova in Rosino's survey (R79), discovered in 1968 September, was found to be spatially coincident with his nova #48 originally seen five years earlier in 1963 September. Three additional eruptions of the nova, now identified as M31N 1963-09c, have been observed since 1968 (Shafter et al. 2015).

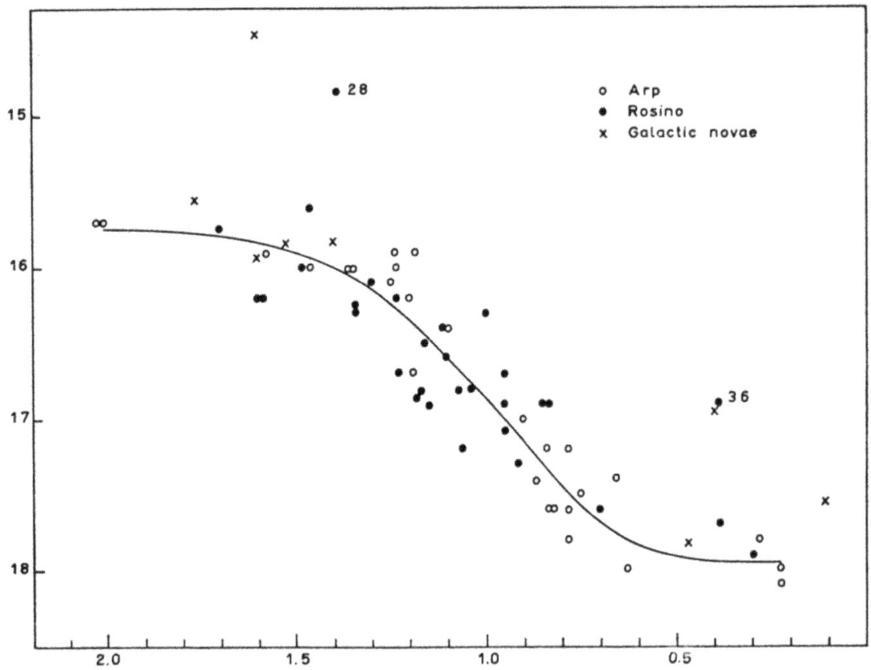

Figure 3.4. The MMRD relation for M31 novae from the Arp (1956) and Rosino (1964) surveys compared with Galactic novae. The ordinate is m_{pg}, with the abscissa given as log (100*d*), with *d* the rate of decline in mag d^{-1}. The particularly luminous Galactic nova is CP Pup. (From Rosino 1964. Reprinted by permission from Springer Nature.)

In addition to the recurrent nova M31N 1963-09c, Rosino discovered another nova candidate that he observed to erupt twice within roughly two years—first in 1966 August (M31N 1966-08a) and then again in 1968 October (M31N 1968-10c). The short recurrence time coupled with the brief duration of the outbursts (a few days) suggested that the object was possibly a Galactic dwarf nova (see Chapter 4) seen in projection against M31. Curiously, no further eruptions of this source have been observed. The nature of the object remained a mystery until just recently when Shafter and associates (Shafter et al. 2017) secured a spectrum of the quiescent counterpart of the transient source and determined it to be a Galactic flare star.

The final phase of the Rosino survey was published in 1989, with the addition of another 52 novae discovered during the period between 1971 and 1986 (Rosino et al. 1989). Over the entire course of the survey a total of 142 novae were discovered in M31, with finding charts and light curves provided for most objects. The survey serves as a major resource for nova studies; in particular, for the future identification of additional recurrent novae.

3.4 Sharov and Alksnis Wide-field Survey

Starting in 1967, yet another photographic survey of novae in M31 was initiated, this time by Russian and Latvian astronomers, Alexander Sergeevich Sharov and

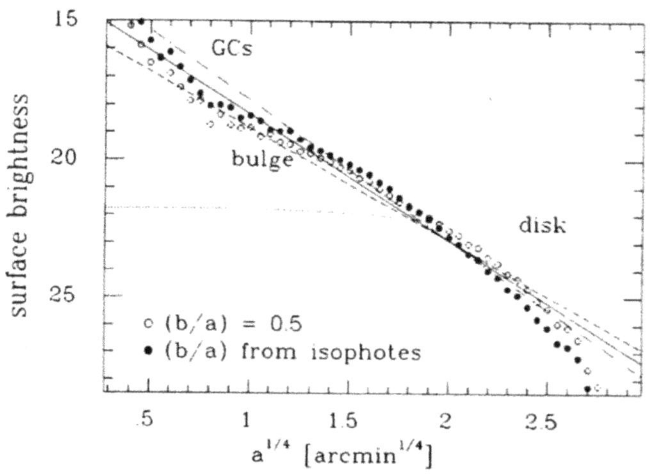

Figure 3.5. The surface density of novae in M31 compared with de Vaucouleurs (1948) light profiles (mag arcsec^{-2}) for the bulge (solid and dashed lines), the globular cluster population (dotted–dashed line), and an exponential model for the disk (dotted line). (Reproduced from Capaccioli et al. 1989. © 1989. The American Astronomical Society. All rights reserved.)

Andrejs Alksnis, using a Schmidt camera on the Radio-Astrophysical Observatory, Latvian SSR Academy of Sciences, and the Maksutov camera of Crimean laboratory of the P. K. Shternberg State Astronomical Institute (Sharov & Alksnis 1989, 1991). Fields with a diameter of 5.5° were obtained with the Schmidt camera, while the Matsukov yielded a 3.6° × 3.6° field. A goal of the survey was to discover novae over a large area of M31 in order to determine whether light curve properties varied as a function of position in the Galaxy. Light curves for a total of 21 novae were produced, with many novae having been previously identified in other surveys. No compelling evidence was found to support the proposition that nova properties varied as a function of distance from the center of M31. Another contribution of the Sharov and Alksnis survey was the realization that several previously discovered nova candidates were, in fact, long-period variables rather than novae.

3.5 The Capaccioli et al. Study

In an important paper, Capaccioli et al. (1989) analyzed the full sample of M31 novae discovered in the major systematic surveys of Hubble (1929) Arp (1956), and Rosino (1964, 1973; Rosino et al. 1989). Perhaps not surprisingly given the sources of the data, the results of the Capaccioli et al. study were mostly in line with those of previous M31 nova studies. Specifically, after correcting for all known observational biases, these authors derived a global nova rate of 28 per year, in very close agreement with the rates of 30 and 29 per year found in the earlier surveys by Hubble and Arp, respectively. In addition, Capaccioli et al. confirmed that the spatial distribution of novae closely follows a de Vaucouleurs (1948) ($r^{1/4}$) bulge light profile, and they were able to place an upper limit of $\theta \equiv \rho_d/\rho_b \sim 0.15$ on the disk-to-bulge nova density. A comparison of the surface density of novae with the surface brightness of M31 novae is shown in Figure 3.5.

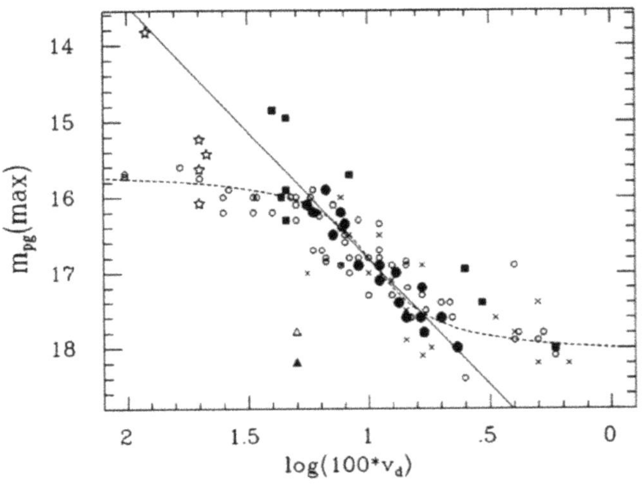

Figure 3.6. The MMRD relation for 105 M31 novae from the combined Arp and Rosino surveys (see Capaccioli et al. 1989 for the key to the different symbols). The solid line is given by Equation (3.1), and the dotted line by Equation (3.2). The discrepant points shown by the filled and open triangles represent Arp's nova #4 and Rosino's nova #48 (= R#79), respectively. The former nova discussed earlier is likely to be heavily extincted, while the latter object is a recurrent nova. (Reproduced from Capaccioli et al. 1989. © 1989. The American Astronomical Society. All rights reserved.)

One of the most important drivers of the Capaccioli et al. study was to carefully re-evaluate the MMRD relation for novae in M31. By the late 1980s a large number of nova light curves had become available, with a total of 105 novae from the Arp (1956) and Rosino (1964, 1973; Rosino et al. 1989) surveys having sufficiently well-sampled light curves to be included in the analysis. Of these, 27 novae had particularly high-quality light curves that were well sampled near maximum light. The MMRD relation from the Capaccioli et al. study is shown in Figure 3.6.

The best linear fit over a rage of $10 < t_2 < 50$ days was given by Capaccioli et al. as:

$$m_{\rm pg}(\max) = (16.88 \pm 0.15) - (3.35 \pm 0.16) \times [\log(100 \times \nu_{\rm d}) - 0.98], \qquad (3.1)$$

where $\nu_{\rm d}$ is the fade rate in mag $\rm d^{-1}$. The fade rate can also be cast in terms of the traditional t_2 parameter, the time (in days) for a nova to fade by 2 mag. After correcting for the color of novae at maximum light, the foreground extinction, and the distance to M31, the relation becomes:[5]

$$M_V(\max) = -12.18 + 3.35 \times \log(t_2). \qquad (3.2)$$

This equation can be compared with the Galactic MMRD relation determined by Cohen (1985):

[5] $(m - M)_{\rm o} = 24.38$ (Freedman et al. 2001), $A_V \simeq 0.20$, $m_{\rm pg} = B - 0.17$ (Arp 1956), $(B - V)_{\max} = 0.23$ (van den Bergh & Younger 1987).

$$M_V(\max) = -10.70 + 2.41 \times \log(t_2). \tag{3.3}$$

The Galactic nova relation is much flatter, with a slope of 2.41 rather than 3.35. As noted by Capaccioli et al. the peak luminosities flatten out for novae with the fastest fade rates, making it clear that the data could not be well fit by a linear relation. They found that a somewhat better fit could be achieved with a non-linear fit of the form:

$$M_V(\max) = -7.89 - 0.81 \times \arctan\left[\frac{1.32 - \log(t_2)}{0.19}\right]. \tag{3.4}$$

This non-linear formulation for the MMRD would remain the default formulation throughout the 1990s.

In summary, the Capaccioli et al. (1989) study provided an excellent summary of what nearly a century of photographic imaging surveys had taught us about novae in M31. Later, in Part II, we will revisit the MMRD relation, and explore what the modern CCD imaging surveys have been able to contribute to our understanding of novae both in M31 and beyond.

References

Arp, H. C. 1956, AJ, 61, 15
Baade, W., & Zwicky, F. 1938, ApJ, 88, 411
Capaccioli, M., della Valle, M., D'Onofrio, M., et al. 1989, AJ, 97, 1622
Cohen, J. G. 1985, ApJ, 292, 90
DeVaucouleurs, G. 1948, AnAp, 11, 247
Freedman, W. L., Madore, B. F., Gibson, B. K., et al. 2001, ApJ, 553, 47
Grubissich, C., & Rosino, L. 1955, Contrib. Asiago #68
Grubissich, C., & Rosino, L. 1957, Contrib. Asiago #77
Grubissich, C., & Rosino, L. 1959, Contrib. Asiago #93
Hubble, E. 1929, ApJ, 69, 103
McLaughlin, D. B. 1945, PASP, 57, 69
Rosino, L. 1964, AnAp, 27, 498
Rosino, L. 1973, A&AS, 9, 347
Rosino, L., Capaccioli, M., D'Onofrio, M., & della Valle, M. 1989, AJ, 97, 83
Shafter, A. W., Henze, M., Rector, T. A., et al. 2015, ApJS, 216, 34
Shafter, A. W., Henze, M., Darnley, M. J., et al. 2017, RNAAS, 1, 44
Shapley, H., & Ames, A. 1932, AnHar, 88, 41
Sharov, A. S., & Alksnis, A. 1989, SvAL, 15, 382
Sharov, A. S., & Alksnis, A. 1991, Ap&SS, 180, 273
van den Bergh, S., & Younger, P. F. 1987, A&AS, 70, 125
Zwicky, F. 1938, ApJ, 88, 529

Chapter 4

A Physical Model for Novae

4.1 The Cataclysmic Variable Stars

The application of photography to the study of variable stars in the middle of the nineteenth century not only facilitated the discovery of large numbers of novae, both in the Milky Way and beyond, it also revealed a new class of eruptive variable star, now referred to as the dwarf novae. U Geminorum, the first of the dwarf novae, was discovered by English astronomer John Russell Hind (1823–1895) on the evening of 1855 December 15 (Hind 1856). The discovery came just seven years after Hind—a well-known observer of asteroids—discovered the first nova of the modern era, V841 Oph (Hind 1848). U Gem reached 9th magnitude at the peak of its eruption, and was initially assumed to be another nova similar to V841 Oph. The peculiar nature of the star wasn't appreciated until subsequent eruptions were observed on 1856 March 26 and 1857 April 7 (Pogson 1857). Continuing observations revealed that U Gem exhibited quasi-periodic eruptions of approximately five magnitudes ($\sim$ 14th mag at quiescence to $\sim$ 9th magnitude in eruption) with a mean inter-outburst interval of about 100 days.

U Gem remained the only variable of its type known for the next 40 years until an eruption of SS Cygni was discovered by one of Pickering's assistants, Louisa D. Wells, in 1896 (Pickering & Fleming 1896). SS Cyg was slightly brighter than U Gem throughout its cycle, alternating between a quiescent magnitude of $\sim$ 12 and an outburst magnitude of $\sim$ 8. By 1940 about 40 similar eruptive variable stars were known, and were collectively referred to after their prototypes as either SS Cyg stars, or alternately, U Gem stars. The brighter and more frequently erupting SS Cyg was observed routinely, and by the 1930s it was probably the most observed variable star of any type.

In addition to the novae and the dwarf novae, another class of irregular variables was recognized in the early twentieth century—the so-called "nova-like" stars. The term nova-like variables referred to stars that varied or "flickered" in brightness by approximately a magnitude or so, on short timescales, and in an aperiodic fashion.

doi:10.1088/2514-3433/ab2c63ch4

Typical members of this class included stars such as AE Aqr, UX UMa, UZ Tau and EX Lup (HV11976). The latter two stars would later be shown to be young stars contracting towards the main sequence, now known as T Tauri stars.[1]

By the middle of the twentieth century, the connection between novae, dwarf novae, and the nova-like variables—stars now collectively referred to as the cataclysmic variable stars—was yet to be appreciated.[2] With regard to novae, our understanding going into the decade of the 1940s was summed up by Hubble in a paper published in 1941, wherein he stated: "Both the galactic system and the neighboring spiral contain several thousand million stars, and both produce several novae per year. There are two possible interpretations of these facts. On the one hand, we may live in an age of novae, at a time in the life history of stellar systems when the stars are in a critical condition and likely to become unstable. On the other hand, the outbursts may be normal episodes in the life history of all stars. On the latter alternative any star might be expected to blow off its cover once in a thousand million years or so. The question is a matter of speculation, and will not be further discussed" (Hubble 1941).

The first insights into the true nature of the nova phenomena began to take shape in the mid-1950s. By then it was becoming increasingly clear that classical novae, along with the U Gem stars and certain nova-like variables (i.e., AE Aqr and UX UMa), were likely all close binary systems. The first clues were provided almost a decade earlier when Roscoe Sanford of the Mount Wilson Observatories made the ground-breaking discovery that the recurrent nova T CrB (1866, 1946) was a binary star with a 230 day orbital period (Sanford 1947, 1949). Sanford secured a total of 29 high-resolution Coudé spectra, which were obtained between 1946 March and 1948 May, yielding the radial velocity curve shown in Figure 4.1.

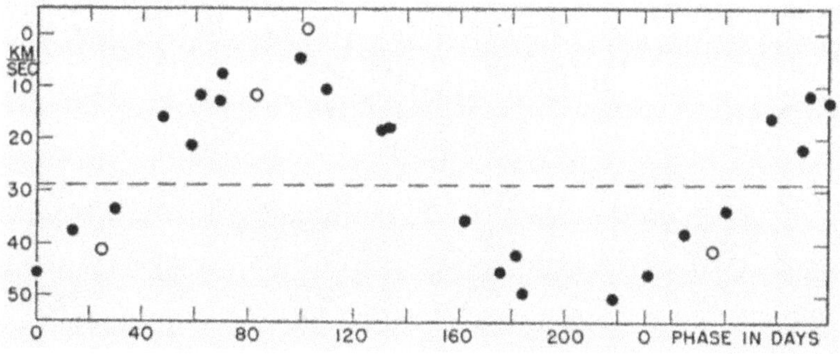

Figure 4.1. Radial velocity curve of the recurrent nova T CrB, establishing that it was a spectroscopic binary with a period of 230.5 days. (Reproduced from Sanford 1949. © 1949. The American Astronomical Society. All rights reserved.)

[1] EX Lup would later be recognized as the prototype of a class of eruptive T Tauri stars dubbed "Exors."

[2] Although R. P. Kraft was the first astronomer to make regular use of the term "cataclysmic variable" to refer to close binary stars consisting of a white dwarf with a low mass companion, he was not the first to use the term. In their book "Variable Stars," published in 1938, Cecilia Payne-Gaposchkin and Sergei Gaposchkin, referred to novae and related stars under the broad term "cataclysmic variables."

In 1954, Alfred H. Joy (1882–1973) published a radial velocity study of the peculiar "flaring" variable star AE Aquarii (Joy 1954). His observations established that the object was a binary star system with an orbital period of just 0.7 days,[3] and mentioned that both the bright and well-known dwarf novae, SS Cyg and RU Peg, had shown radial velocity variations consistent with a binary nature.

The real turning point, however, came when Merle F. Walker of the Lick Observatory obtained time-resolved photometric observations of the well-known classical nova DQ Her (nova Her 1934), which revealed the object to be an eclipsing binary with a 4 h 39 m period (see Figure 4.2; Walker 1954). At the time this was one of the shortest orbital period known for any binary star system, second only to that of the nova-like variable UX UMa—a well-known but poorly understood system whose relation to the novae and dwarf novae was not yet fully appreciated. In Walker's words: "The light-curve of DQ Her is very similar to that of the other known eclipsing binary of extremely short period, UX UMa. It would, of course, be premature to conclude that UX UMa is in any way related to the novae."

The first model for the basic structure of the cataclysmic variables came in 1956 in an important paper by Crawford & Kraft (1956) on the peculiar variable AE Aquarii. They argued that AE Aqr was a semi-detached binary system with mass being transferred through the inner Lagrangian point to a compact blue companion via an accretion disk as illustrated in Figure 4.3. Although the relationship of AE

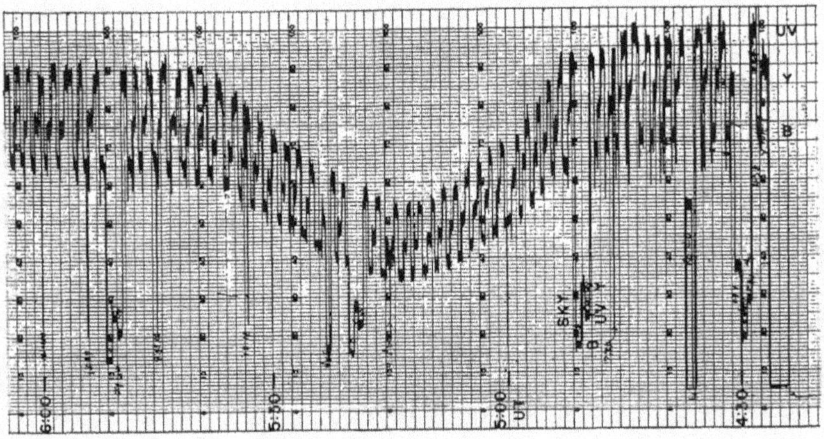

Figure 4.2. A section of the Brown recorder sheet from Merle Walker's 1954 August 1 observations of DQ Her (nova Her 1934) when the first eclipse was discovered (time on abscissa and relative intensity on ordinate; Walker 1954). Note the short-period (71 s) fluctuations present in the UV, blue, and yellow light curves that would later be associated with the white dwarf's rotation. (© 1954. The Astronomical Society of the Pacific. All rights reserved. Printed in U.S.A.)

[3] Actually, there was a one cycle per day alias in the period determination, and the true orbital period of AE Aqr is 0.411 days (Walker 1965).

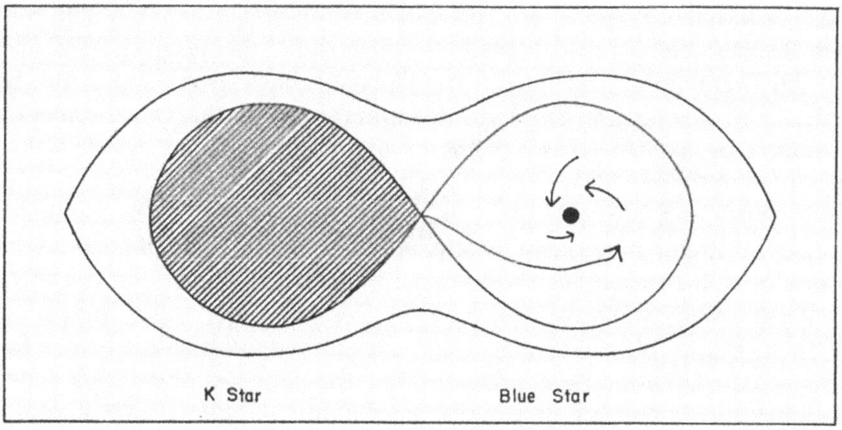

Figure 4.3. A schematic representation of the cataclysmic binary AE Aqr showing a Roche-lobe-filling K-star transferring mass to a compact blue star through the inner Lagrangian point. (Reproduced from Crawford & Kraft 1956.)

Aqr to the novae was not firmly established as yet, this was the first time that mass transfer in a semi-detached binary system was invoked to explain the properties of a cataclysmic variable star.

In a series of seminal papers (*Binary Stars among Cataclysmic Variables*) starting in the 1960s Robert P. Kraft (1927–2015), of the Mount Wilson and Palomar Observatories and later the Lick Observatory, was able to establish beyond reasonable doubt that the cataclysmic variable stars were all close binary systems (Kraft 1962; Kraft et al. 1962; Kraft 1964a, 1964b; Krzeminski & Kraft 1964; Kraft et al. 1969). In the first paper of the series, Kraft (1962) established the binary nature and orbital periods for five dwarf novae (SS Cyg, U Gem, RX And, RU Peg, and SS Aur), EY Cyg was shown to have a composite spectrum consistent with a binary nature, and Z Cam, prototype of the class of dwarf novae showing "standstills,[4]" exhibited radial velocity variations of yet undetermined period. In the five cases where the orbital period was determined, it was found to be less than nine hours, establishing the compact nature of these binaries.

Two years later, in the third paper of the series, Kraft (1964a) presented spectroscopic observations of ten "old novae." In addition to four systems with previously determined periods (T CrB, DQ Her, T Aur, and WZ Sge), Kraft determined orbital periods for an additional four systems: GK Per (Nova Per 1901), V603 Aql (Nova Aql 1918), and the recurrent nova V1017 Sgr (Nova Sgr 1901, 1919). Two additional novae (V841 Oph and DI Lac) did not show significant radial velocity variations and were assumed to have orbital inclinations that were nearly "pole-on."

The spectra of the cataclysmic variable stars were typically dominated by strong and broad emission lines of hydrogen and helium. In systems with orbital periods in

[4] Standstills are periods of near constant luminosity where the dwarf nova remains in outburst at a level just below peak brightness.

excess of ~6 h, absorption features characteristic of a late-type red star were also observed. Radial-velocity measurements established that the absorption features from the red star were anti-phased with the emission lines associated with the blue component. The observations also indicated that the red star had a radius comparable to the semimajor axis of its orbit, and was transferring mass through the inner Lagrangian point to the blue component, which Kraft (1964a) described as a "a semi-degenerate object in a late stage of stellar evolution." As was the case with AE Aqr, conservation of angular momentum required that the transferred material initially form a ring or disk surrounding the compact blue star. It was in this disk of rapidly-rotating gas (typically some thousands of km s^{-1}) where the broad emission features seen in the spectra of old novae and dwarf novae alike were believed to be formed.

By the mid 1960s, despite confidence that the cataclysmic variable stars were compact binary stars, the nature of the nova eruption and its possible relationship to the dwarf nova eruption was far from resolved. For example, it was still unclear if the red or the blue component was the seat of the eruptions. In the case of the well-known dwarf nova U Gem, an influential study by Krzeminski (1965) who studied the evolution of eclipse depths through outburst seemed to establish that the eruption was associated with the red component. It was clear at the time, however, that the nova eruption released a total energy of order 10^{45} erg, while the energy release from dwarf nova was some five orders of magnitude less.

The outbursts of dwarf novae would eventually be understood as arising from the liberation of gravitational potential energy in the accretion disk as material was quasi-periodically accreted onto the white dwarf. Initially, it was unclear if the mass accretion was modulated by instabilities in the secondary star (e.g., Bath 1976) or in the accretion disk itself (e.g., see Pringle 1981). The topic was the subject of a great deal of consternation in cataclysmic variable community in the early 1980s, with the Roche-lobe overflow advocates and their disk instability counterparts often engaging in spirited debates at various conferences and workshops. Eventually, the issue was settled, and it became clear that accretion disk instabilities were responsible for dwarf nova eruptions (Meyer & Meyer-Hofmeister 1981; Cannizzo et al. 1992; Cannizzo & Kaitchuck 1992).

An early explanation for the nova phenomena was based upon a model proposed by Evry Schatzman in 1958 (Schatzman 1958). In this model, long-period non-radial oscillations are triggered in the blue component, and amplified by a resonance with the orbital period. As the star evolved, the oscillations can eventually result in explosive ejection of material along the polar caps and several belts or zones. Although the Schatzman model required the nova progenitor to be a close binary system, it did not require mass transfer from the red to the blue component as was observed in the cataclysmic variable stars.

4.2 Thermonuclear Runaway Models

By the time it became clear that nova progenitors were all close binary systems, the idea that classical nova eruptions might be powered by thermonuclear processes had been around for some time (e.g., Schatzman 1949; Gurevitch & Lebedinsky 1957).

Initially, Kraft adopted a model for the nova eruption that involved a thermonuclear runaway (TNR) in the blue component, triggered by accretion of hydrogen from the red companion. However, by 1964 he had given up on the viability of such a model (Kraft 1964a), concluding that the heat generated in the burning region would be conducted to the entire degenerate core of the blue component. In particular, a decade earlier, Mestel (1952) had argued that in such a case the ensuing TNR would engulf the entire star, with the energy released being comparable to that of a supernova.

By the end of the 1960s, the theoretical picture changed fundamentally when the first models exploring the accretion of hydrogen onto the surfaces of white dwarfs were developed by Giannone & Weigert (1967) and by Rose (1968). Shortly thereafter, in a series of seminal papers, Sumner Starrfield and collaborators developed the basic model for the nova eruption as a thermonuclear runaway on the surface of an accreting white dwarf (Starrfield 1971; Starrfield et al. 1972, 1974a, 1974b).

Initially, Starrfield (1971) computed TNR models on the surfaces of low mass (0.12 and 0.5 $M_\odot$) He white dwarfs. The lower mass was chosen to match that assumed at the time for the blue component of nova DQ Her (Kraft 1963). The result was that a thermonuclear runaway on the surface of a 0.12 $M_\odot$ white dwarf did not produce the required 10^{45} erg of energy observed in the outburst. In retrospect, this result was not surprising given that the mass assumed in the model was much lower than the best estimate at the time for the mass of the white dwarf in DQ Her. In subsequent papers, Starrfield et al. (1972, 1974a, 1974b) computed models seated on higher mass white dwarfs and demonstrated the importance of the CNO abundances on the subsequent energy released in the TNR. Although the proton–proton (p–p) chain is initially important during the mass accretion phase, at the temperatures expected at the base of the accreted hydrogen-rich layer at the onset of the TNR, the energy production from the CNO cycle (see Figure 4.4) dominates over the p–p chain. A summary of the state of our understanding of

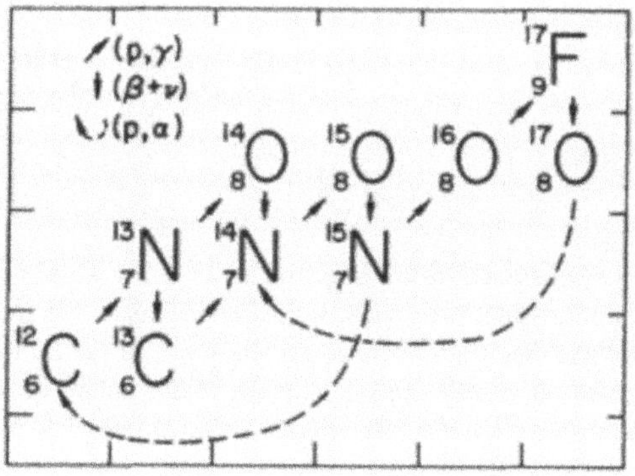

(a)

Figure 4.4. The CNO reaction chain relevant to novae. Proton captures are followed by either $\beta+$ decays, or alpha decays. (Reproduced from Starrfield et al. 1972. © 1972. The American Astronomical Society. All rights reserved.)

novae as the decade of the 1970s was winding down can be found in an early review by Gallagher & Starrfield (1978).

Over the nearly half century since the TNR models were first proposed, considerable progress has been made in understanding the eruption process. In the early 1980s Masayuki Fujimoto and Ken'ichi Nomoto studied the consequences of accretion onto white dwarf stars—both in relation to nova outbursts, and as a possible triggering mechanism for Type Ia supernovae (Fujimoto 1982; Fujimoto & Taam 1982; Nomoto 1982). Both investigations suggested that depending on the rate, the accretion would result in the formation of a red giant atmosphere (high rates $\gtrsim 10^{-6}\ M_{\odot}\ yr^{-1}$), steady nuclear burning on the WD surface (intermediate rates of a few $\times\ 10^{-7}\ M_{\odot}\ yr^{-1}$), or hydrogen shell flashes (novae) for accretion rates $\lesssim 10^{-7}\ M_{\odot}\ yr^{-1}$, with a weak dependence on the white dwarf mass. The plots shown in Figure 4.5 nicely illustrate how the accreted mass required to initiate a TNR (the ignition mass) depends on fundamental system parameters such as WD mass and accretion rate.

The 1990s saw the first attempts to model a nova progenitor system through several eruption cycles. Prialnik & Kovetz (1995) produced a grid of nova models that explored how the observed properties of the outburst (luminosity, ejection

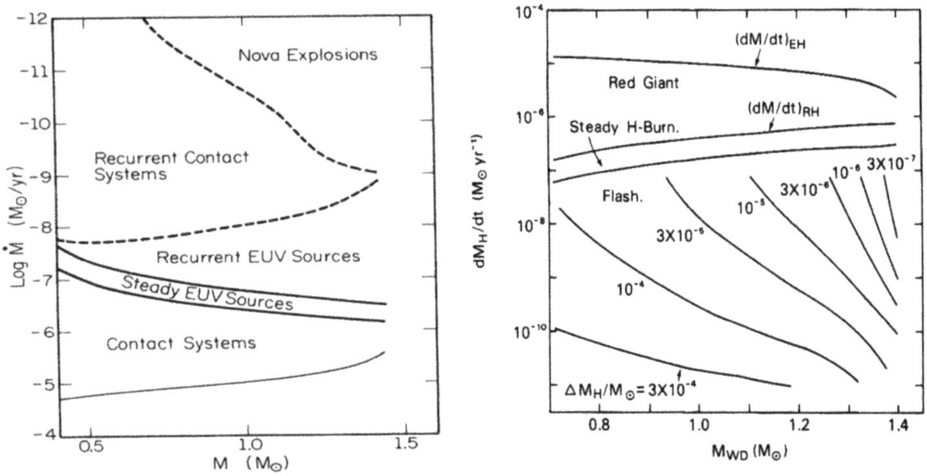

(a) (b)

Figure 4.5. (a) Characteristics of hydrogen shell burning on an accreting white dwarf as a function of its mass and accretion rate (from Fujimoto 1982). The thick solid lines delineate the stable burning regime without expansion, while the thin solid line denotes the critical accretion rate at the Eddington luminosity. Below the lower dashed line shell flashes lead to a shell expansion of less than a solar radius, and above the top dashed line shell flashes cause a high expansion velocity. Between two dashed lines shell flashes lead to large expansion at relative low velocity. (b) A similar plot produced by Nomoto (1982), later to become known as the "Nomoto Diagram." For sufficiently low accretion rates, the burning is unstable and leads to hydrogen shell flashes (i.e., nova eruptions). The ignition mass is lowest for higher WD masses and accretion rates. Such systems have the shortest periods between successive eruptions, and will be most likely to be recognized as recurrent novae. (Reproduced from Fujimoto 1982 and Nomoto 1982. © 1982. The American Astronomical Society. All rights reserved.)

velocity, fade rate, and recurrence time) depends on basic system parameters (white dwarf mass, temperature and accretion rate). More sophisticated models continue to be developed, and an excellent discussion of recent work can be found in the 2016 review by Starrfield et al. (2016).

4.3 Spectroscopic Classes of Novae

Spectroscopic observations of novae go all the way back to 1866 when William Huggins and chemist W. Allen Miller undertook a pioneering study of the recurrent nova T Corona Borealis (Huggins 1866; Huggins & Miller 1866). Figure 4.6 shows a reproduction of a spectrum taken on the night of 1866 May 16, four days after the discovery of the nova.[5] Huggins and Miller describe the spectrum as consisting of two principal components. In their words "When the spectroscope was placed on the telescope, the light of this new star formed a spectrum unlike that of any celestial body which we have hitherto examined. The light of the star is compound, and has emanated from two different sources. Each light forms its own spectrum. In the instrument these spectra appear superposed." They were of course referring to the absorption and emission components that are commonly seen in the spectra of novae shortly after eruption. The bright lines seen in Figure 4.6 and marked as Fraunhofer lines C and F correspond to the Balmer lines Hα and Hβ, while the absorption feature marked as Fraunhofer D refers to the well-known sodium D doublet at λ5890, 5896 Å.

The evolution of nova spectra have traditionally been described following a nomenclature that goes back to a paper published by McLaughlin in 1937 (McLaughlin 1937), and refined in subsequent years (McLaughlin 1943; 1960). McLaughlin divided nova spectra into absorption and emission systems, and proposed the following stages for each: (1) Pre-maximum, (2) Principal, (3) Diffuse enhanced, (4) Orion, and (5) Nebular. The pre-maximum spectrum is not often observed, but when observations have been made they show a blue continuum, similar to stars of spectral type B, and contain weak absorption features that are

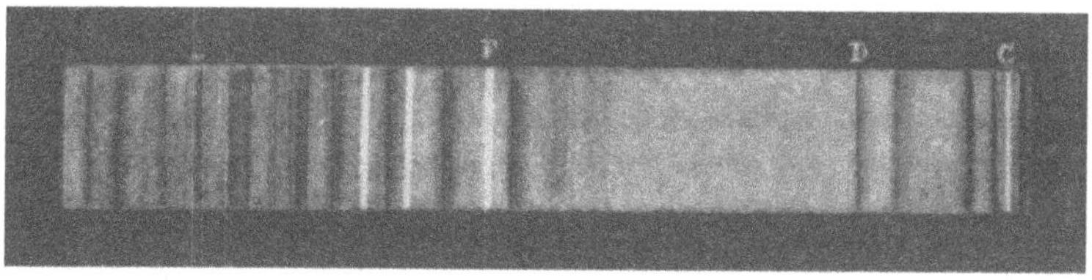

Figure 4.6. The first published spectrum of a nova: T CrB (Nova Corona Borealis 1866). This drawing of the spectrum shows numerous absorption features including the sodium D lines, and several Balmer emission lines (Hβ marked F and Hα marked C, after their Fraunhofer designations). (Reproduced from Huggins & Miller 1866 with permission from The Royal Society of London.)

[5] Although the discovery of T CrB on the night of 1866 May 12 is generally credited to amateur astronomer John Birmingham (e.g., Lynn 1866a), there is some controversy as to the date the object was first seen (e.g., see Hind et al. 1866; Lynn 1866b).

strongly displaced to the blue. The principal spectrum is characteristic of maximum light, and strongly resembles that of a supergiant star with spectral class A or F. One difference noted by McLaughlin is that nova spectra often contain neutral lines of neutral carbon and oxygen. Emission components begin to emerge during the principal stage and always consist of hydrogen lines, often accompanied by Fe II and the sodium D lines. The combination of emission and blueshifted absorption lines result in the P Cygni line profiles[6] that are characteristic of novae at maximum light. During the diffuse-enhanced stage that occurs days to weeks after maximum light (about 2 mag below maximum light), the absorption lines (typically H, Na I, Ca II, Fe II) become wide and diffuse shifting further to the violet, the continuum weakens, the emission lines begin to strengthen. The Orion stage is also characterized by diffuse absorptions with N II, O II and He I dominating. The Orion emission spectrum often consists of a number of diffuse lines of N II and N III features, including the so-called "λ4640" emission feature due to N III. The final stage of nova spectral development is the Nebular stage where the excitation of the spectrum increases as the nova fades. Forbidden lines of O I (e.g., λ6300, 6364 Å) and N II (λ5755 Å) emerge early, followed later by [O III] λ4959, 5007 Å and [Ne III] λ3869, 3968 Å emissions in some systems after the novae have faded four or more magnitudes below maximum light. Coronal lines such as [Fe X] λ6374 Å and [Fe XIV] λ5303 Å also appear late in the evolution of some novae. The stages of spectral development formulated by McLaughlin have become widely adopted and their use continues to the present day. A wealth of data on the spectroscopic characteristics of novae from the first half of the twentieth century can be found in Cecilia Payne Gaposchkin's classic book, "The Galactic Novae" (Payne-Gaposchkin 1957).

The next major development in the classification of nova spectra came in an influential paper (*The Formation of Novae Spectra*) published in the early 1990s, when Robert Williams suggested that nova spectra could be classified into one of three principal types: the Fe II, He/N, and Hybrid classes (Williams 1992). His classification scheme was empirical in nature and was based on the general spectroscopic characteristics of optical nova spectra taken within the first few weeks of eruption (i.e., in the so-called Diffuse Enhanced phase, before the nova entered the nebular phase). The most common class, the Fe II novae, which make up ~ 80% of Galactic novae, is characterized by prominent Balmer and Fe II emission lines that show P Cygni profiles early in the outburst. Typical examples of novae in the Fe II class are shown in Figure 4.7.

Novae of the He/N class, which comprise roughly 15%–20% of Galactic systems differ markedly. In addition to the Balmer lines, He/N novae are characterized by higher excitation lines of helium and nitrogen—typically He I λ4922, λ5015, λ5876, λ6678, and the Bowen blend of NIII, CIII at 4640, 5001 Å. The lines are characteristically broad (FWHM >2500 km s^{-1}), showing little or no evidence for P Cygni profiles. The spectra are usually devoid of Fe II features. Examples of

[6] A P Cygni profile refers to a line profile that consists of an emission component flanked by a blueshifted absorption feature. The profile results from outflow of gas from a star and is named after the prototype P Cygni.

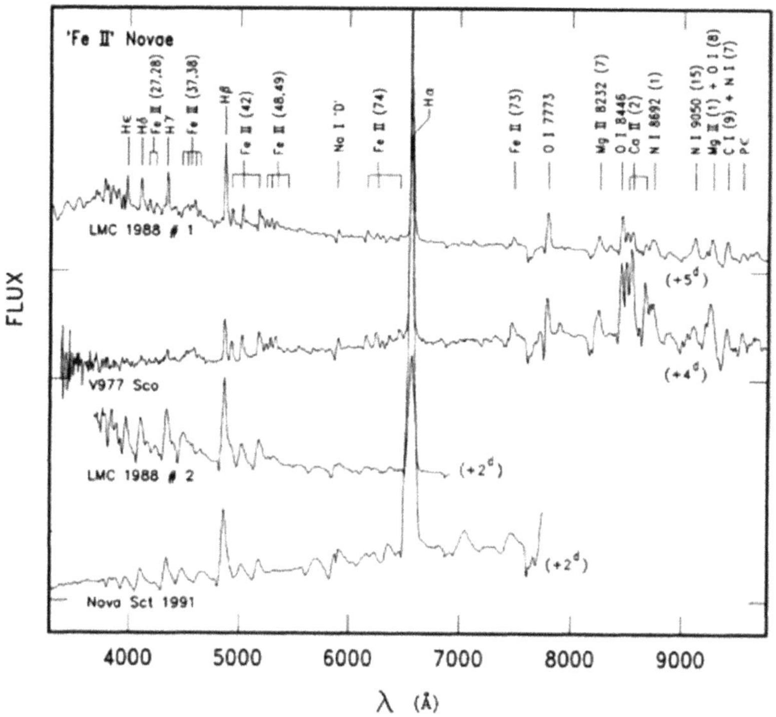

Figure 4.7. Sample spectra of Fe II novae. Numerous Fe II emission features are evident along with P Cygni profiles, which are most conspicuous for the Hβ line. (Reproduced from Williams 1992. © 1992. The American Astronomical Society. All rights reserved.)

He/N nova spectra are shown in Figure 4.8. Finally, Williams describes a "Hybrid" nova class (making up ~5% of the total) that shares some characteristics of both of the principal classes. The spectrum shortly after eruption (within the first week) displays rather broad (FWHM ~2500 km s^{-1}) Balmer and Fe II emission lines that later transition to one characteristic of the He/N class.

The underlying physical processes that give rise to the different spectroscopic classes have not been fully modeled. However, in his 1992 paper and more recently (Williams 2012), Williams has given a qualitative description of the conditions that likely underlie the formation of spectra across the differing classes. During the initial stages of a nova outburst, a discrete shell of material is ejected from the white dwarf's surface at high velocity (>3000 km s^{-1}). The relatively high expansion velocity causes the shell to rapidly become optically thin, producing emission-line profiles with a broad, flat-topped or castellated form. In most novae, residual nuclear burning on the surface of the white dwarf results in additional mass loss via a more slowly expanding wind. Emission-line profiles produced in the comparatively slowly expanding and dense wind are expected to be relatively narrow, and exhibit P-Cygni profiles. The character of the overall nova spectrum depends on which of these two emission components dominates. The He/N novae are thought to originate

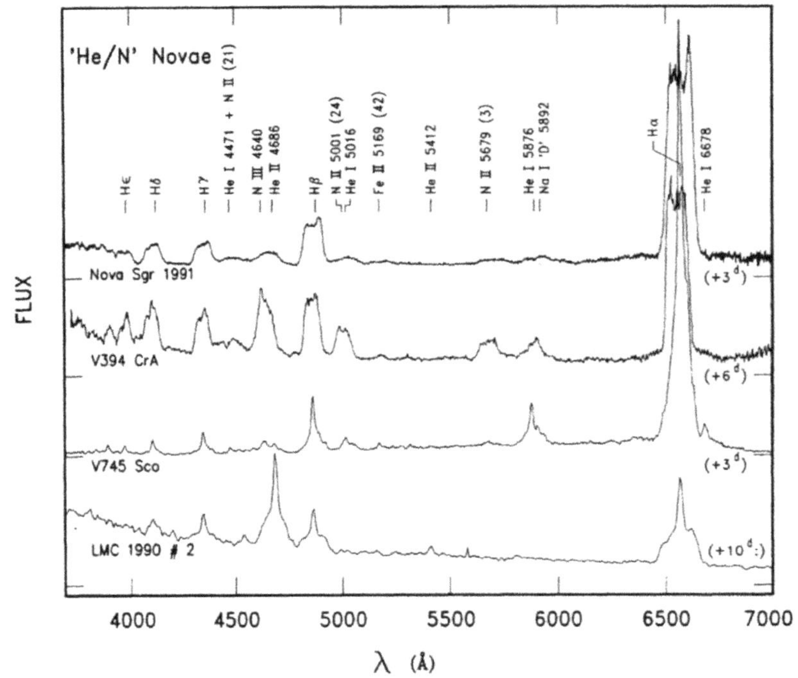

Figure 4.8. Examples of He/N nova spectra. Note the absence of Fe II lines coupled with the strong and broad Balmer, He and N emission features. (Reproduced from Williams 1992. © 1992. The American Astronomical Society. All rights reserved.)

on relatively massive WDs, which require a smaller accreted envelope to achieve a thermonuclear runaway, and thus produce a smaller ejected mass. Such systems should have little residual nuclear burning, a weak wind mass loss component, and thus produce a spectrum dominated by the emission from the rapidly expanding shell. On the other hand, novae with larger accreted envelopes, and ongoing wind mass loss will produce spectra with narrow lines and P Cygni lines characteristic of the Fe II novae.

The relationship between fundamental binary parameters (e.g., WD mass and accretion rate) and the character of the resulting nova spectrum and its evolution is not well understood. Key questions are: Why are there two dominant spectroscopic classes rather than a continuum of spectral variability? Do all novae evolve through the spectroscopic classes, but at differing rates? Although our understanding is incomplete, it seems clear that time-resolved spectroscopic observations of novae will be required to fully elucidate the physics of nova eruptions.

References

Bath, G. T. 1976, in IAU Symp. 73, Structure and Evolution of Close Binary Systems, ed. P. Eggleton, & S. Mitton (Dordrecht: Reidel), 173

Cannizzo, J. K., Ghosh, P., & Wheeler, J. C. 1992, ApJ, 260, 83

Cannizzo, J. K., & Kaitchuck, R. H. 1992, SciAm, 226, 92
Crawford, J. A., & Kraft, R. P. 1956, ApJ, 123, 44
Fujimoto, M. Y. 1982, ApJ, 257, 767
Fujimoto, M. Y., & Taam, R. E. 1982, ApJ, 260, 249
Gallagher, J. S., & Starrfield, S. 1978, ARA&A, 16, 171
Giannone, P., & Weigert, A. 1967, ZA, 67, 41
Gurevitch, L. E., & Lebedinsky, A. I. 1957, in IAU Symp. 3, Non-stable Stars, ed. G. H. Herbig (Cambridge: Cambridge University Press), 77
Hind, J. R. 1848, MNRAS, 8, 146
Hind, J. R. 1856, MNRAS, 16, 56
Hind, J. R., Bird, F., & Lynn, W. T. 1866, AReg, 4, 229
Hubble, E. 1941, PASP, 53, 141
Huggins, W., & Miller, W. A. 1866, RSPS, 15, 146
Huggins, W. 1866, MNRAS, 26, 275
Joy, A. H. 1954, ApJ, 120, 377
Kraft, R. P. 1962, ApJ, 135, 408
Kraft, R. P., Mathews, J., & Greenstein, J. L. 1962, ApJ, 136, 312
Kraft, R. P. 1963, AdA&A, 2, 43
Kraft, R. P. 1964a, ApJ, 139, 457
Kraft, R. P. 1964b, ASPL, 9, 137
Kraft, R. P., Krzeminski, W., & Mumford, G. S. 1969, ApJ, 158, 589
Krzeminski, W., & Kraft, R. P. 1964, ApJ, 140, 921
Krzeminski, W. 1965, ApJ, 142, 1051
Lynn, W. J. 1866a, AN, 67, 197
Lynn, W. J. 1866b, AReg, 4, 310
McLaughlin, D. B. 1937, ApJ, 85, 362
McLaughlin, D. B. 1943, Publications of the Observatory of the University of Michigan (Vol. 8; Ann Arbor, MI: University of Michigan Press), 149
McLaughlin, D. B. 1960, in Stellar Atmospheres, ed. J. L. Greenstein (Chicago, IL: University of Chicago Press), 585
Meyer, F., & Meyer-Hofmeister, E. 1981, A&A, 104, L10
Mestel, L. 1952, MNRAS, 112, 598
Nomoto, K. 1982, ApJ, 253, 798
Payne-Gaposchkin, C. 1957, The Galactic Novae (Amsterdam: North-Holland)
Pickering, E. C., & Fleming, W. 1896, ApJ, 4, 369
Pogson, N. 1857, MNRAS, 17, 200
Prialnik, D., & Kovetz, A. 1995, ApJ, 445, 789
Pringle, J. E. 1981, ARA&A, 19, 137
Rose, W. K. 1968, ApJ, 152, 245
Sanford, R. F. 1947, PASP, 59, 87
Sanford, R. F. 1949, ApJ, 109, 81
Schatzman, E. 1949, AnAp, 12, 281
Schatzman, E. 1958, AnAp, 21, 1
Starrfield, S. 1971, MNRAS, 152, 307
Starrfield, S., Truran, J. W., Sparks, W. M., & Kutter, G. S. 1972, ApJ, 176, 169

Starrfield, S., Sparks, W. M., & Truran, J. W. 1974a, ApJS, 28, 247
Starrfield, S., Sparks, W. M., & Truran, J. W. 1974b, ApJ, 192, 647
Starrfield, S., Iliadis, C., & Hix, W. R. 2016, PASP, 128, 1001
Walker, M. F. 1954, PASP, 66, 230
Walker, M. F. 1965, S&T, 29, 23
Williams, R. E. 1992, AJ, 104, 725
Williams, R. E. 2012, AJ, 144, 98

Part II

The Modern Era

Chapter 5

Nova Rates and Populations in the Milky Way

In the mid 1980s, Hilmar W. Duerbeck (1948–2012), the twentieth century's most recognized chronicler of nova outbursts, was focusing his attention on the population of novae in the Milky Way. In a truly monumental work, Duerbeck produced a detailed compendium of all known Galactic nova candidates from the first suspected nova, CK Vul in 1670[1] up through 1986 (Duerbeck 1987). The catalog summarizes the photometric properties such as the peak magnitude, the magnitude at minimum, if known, and the fade rate (t_3 time) for a total of 277 objects. Of these, approximately 50% have observations sufficient to assure their nova classification, with another 28% almost certainly novae. The remaining 22% include some poorly observed objects such as symbiotic stars, X-ray novae, (long recurrence time) dwarf novae, etc. Since Duerbeck's catalog was published, the number of Galactic novae known has increased by more than 50 percent. A comprehensive list of the nearly 500 nova candidates reported in the Milky Way up through the present is maintained online by William Gray (https://projectpluto.com/galnovae/galnovae.htm). A small, but significant, fraction of these Milky Way candidates are based on limited data, and most have not been spectroscopically confirmed. As a result, a significant number of these detections may be objects other than classical novae (e.g., dwarf novae, flare stars, or errant observations).

5.1 Two Nova Populations?

Despite the complications introduced by the effects of interstellar dust concentrated near the Galactic plane, which bias the observed nova sample to nearby objects, Duerbeck (1990) was able to arrive at several tentative conclusions regarding the

[1] CK Vul was discovered in eruption on 1670 June 20 by Voituret Anthelme, a French Carthusian monk and astronomer. Recent work on the nova remnant suggest that the object was likely not a classical nova, but rather a so-called "red nova," which are thought to result either from the merger of two main-sequence stars (Kamiński et al. 2015), or perhaps a main-sequence star and a white dwarf (Eyres et al. 2018).

doi:10.1088/2514-3433/ab2c63ch5

properties of Galactic novae. His principal result was that Galactic novae could be divided into two distinct populations: a disk population characteristic of novae in the solar neighborhood with a mean scale height of ~150 pc, and a bulge population consisting of novae concentrated towards the Galactic center having a larger average scale height of ~ 500 pc. Novae that are members of the disk population are purported to harbor more massive white dwarfs that result in generally faster outburst development (i.e., shorter t_3 times) compared with their bulge nova counterparts (e.g., Della Valle et al. 1992).

Figure 5.1 shows how the number counts of novae vary as a function of apparent magnitude. Consistent with his two-population idea, Duerbeck noticed that the number counts appear to follow a disk distribution ($N[\mathrm{m}] \propto 10^{0.4m}$) for novae brighter than $m \sim 3$, before incompleteness sets in causing the cumulative number of novae to decrease with increasing magnitude. However, at $m \sim 5$ the number counts begin to increase again, consistent with what is expected from the contribution of a distinct bulge nova population. Figure 5.2 shows the distribution of Duerbeck's

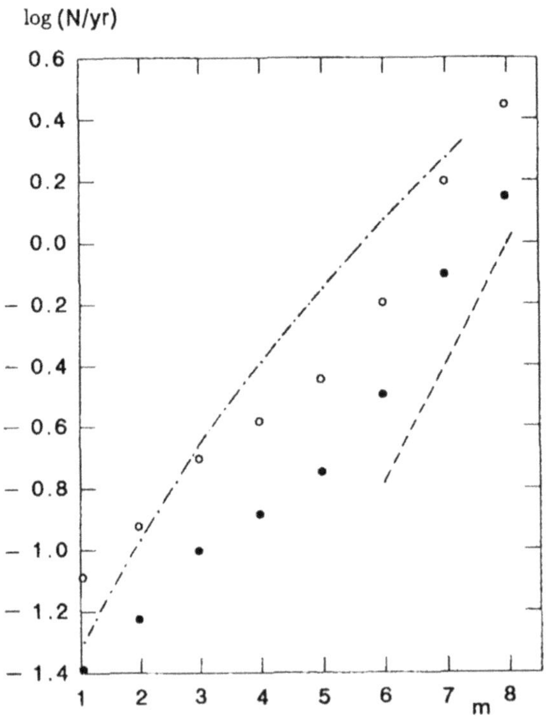

Figure 5.1. The observed number counts of Galactic novae as a function of apparent magnitude. The filled circles represent the observed rates and the open circles show double the observed rate (both corrected for seasonal incompleteness). The dashed–dotted line shows the model number counts for a disk population, while the dashed line shows that expected for a bulge population. It appears that down to $m \sim 5$ we observe mostly disk novae becoming increasingly more incomplete towards fainter magnitudes, with the bulge novae dominating as expected at fainter magnitudes. (From Duerbeck 1990; Springer Science+Business Media B.V. With permission of Springer.)

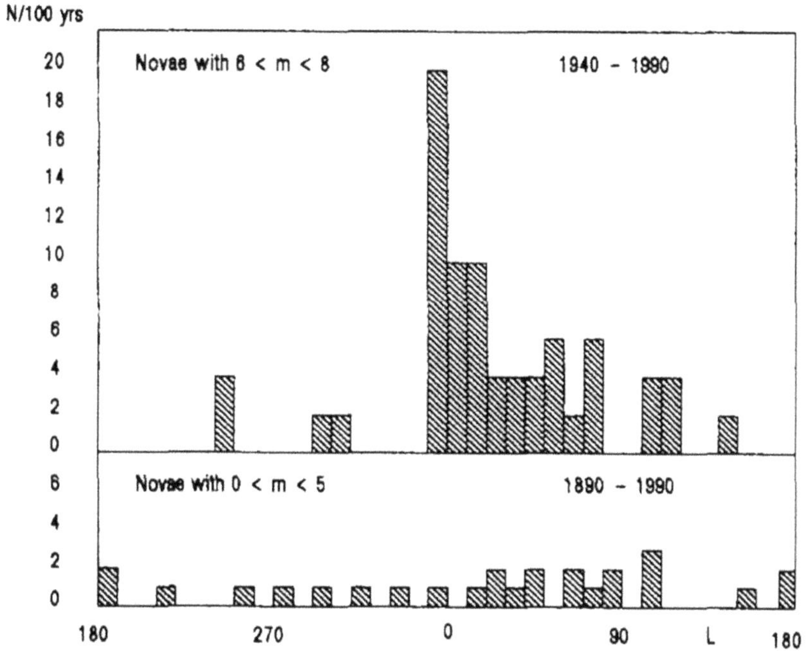

Figure 5.2. The distribution of novae in Galactic longitude. The brighter (presumably disk) novae in the solar neighborhood are evenly distributed in longitude, but the fainter novae show a concentration towards the Galactic center. The asymmetry about longitude 0 is explained by the historically more complete monitoring of the northern sky. (From Duerbeck 1990; Springer Science+Business Media B.V. With permission of Springer.)

Table 5.1. Nova Speed Class in Different Stellar Systems

Location	Very Fast	Fast	Moderately fast	Slow
LMC	56%	44%	0%	0%
Galactic disc	38%	33%	15%	5%
M31	17%	44%	31%	7%
Galactic bulge	24%	30%	36%	9%

Note. Adapted from Duerbeck (1990).

nova sample in Galactic longitude. Clearly, the brighter novae, which are generally nearby and part of the Galactic disk, are more uniformly distributed, while the fainter and generally more distant novae are preferentially located in the direction of the Galactic bulge. These results, while demonstrating that Galactic novae are found in both the bulge and disk of the Galaxy, do not provide evidence that nova properties differ between the populations. That evidence came from Duerbeck's comparison of nova speed classes—Very Fast ($t_2 < 10\,d$), Fast ($11\,d < t_2 < 25\,d$), Moderately Fast ($26\,d < t_2 < 80\,d$), Slow ($81\,d < t_2 < 150\,d$)—in different stellar environments (see Table 5.1).

The Galactic data do indeed show some evidence that the slower developing novae are preferentially associated with the bulge, however it is difficult to assess the statistical significance of this result given that the absolute numbers of novae studied were not specified. In contrast to the Galactic data, some 60% of the M31 novae studied by Duerbeck belong to the "Very Fast" or "Fast" groups. This is somewhat surprising assuming Duerbeck's two-population hypothesis considering that a large fraction of M31 novae appear to be associated with the bulge of that galaxy (Ciardullo et al. 1987; Shafter & Irby 2001; Darnley et al. 2006).

Perhaps the strongest evidence for the two-nova-population concept was provided by Duerbeck's comparison of Galactic (and M31) nova speed class distribution with that observed in the Large Magellanic Cloud (LMC). At the time of Duerbeck's analysis light curve data sufficient to determine speed classes were only available for approximately a dozen LMC novae. Nevertheless, it is remarkable that all of the novae considered by Duerbeck were either classified as "Very Fast" or "Fast." It is unclear which LMC novae Duerbeck used in his analysis, however, as a check of the literature reveals that there were at least two objects, LMCN 1926-09a and LMCN 1948-12a, with reported t_3 times of 200 and 100 days, respectively, that appeared to be rather slow novae (Buscombe & de Vaucouleurs 1955). All told, it seems fair to conclude that circa 1990 the evidence for two nova populations was not particularly compelling.

On the heels of the pioneering Duerbeck study, Massimo Della Valle and collaborators were pursuing a study to determine whether the absolute magnitudes and decline rates of Galactic novae varied as a function of height (z) above the Galactic plane. Because accurate distances are required for such a study, Della Valle et al. restricted their analysis to a "fiducial" sample of 19 novae with distances determined from expansion parallax (Della Valle et al. 1992). The distribution of rates of decline for the sample revealed a gap centered on t_2 = 13 d. Seven of the 19 novae with $t_2 < 13$ d were were placed in a "fast" nova group, with the remaining 12 novae constituting the "slow" group. When the z values of the nova sample were computed, six out of seven novae in the fast group were found to lie within 100 pc of the Galactic plane, with nine of the 12 slow novae having z distances between 100 and 1000 pc (see Figure 5.3). In addition, Della Valle et al. found that the average absolute magnitudes at maximum light differed significantly between the groups with the fast novae characterized by $\langle M_V \rangle = -9.31 \pm 0.21$, and the slow novae by $\langle M_V \rangle = -7.53 \pm 0.39$. The authors reported that a Kolomogorov–Smirnov (KS) test supported the hypothesis of two distinct stellar populations characterized by different z values and different absolute magnitudes at maximum light to a high level of significance ($P < 0.01$ that the samples were drawn from the same parent distribution).

5.1.1 Spectroscopic Classes of Galactic Novae

Another approach to exploring the populations of Galactic novae is to consider their spectroscopic classifications, which were introduced in Chapter 4. Specifically, given the results from the Duerbeck and Della Valle et al. studies just described, it makes sense to ask whether the two principal spectroscopic classes of novae (Fe II and He/N) also vary as a function of spatial position within the Galaxy. Such a study was

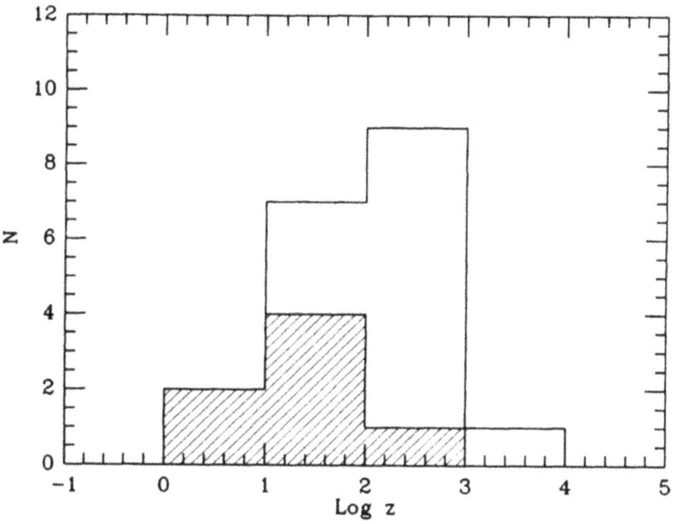

Figure 5.3. The height above the Galactic plane for a sample of fast (shaded) and slow (unshaded) novae. (Reproduced with permission from Della Valle et al. 1992, © ESO.)

pursued by Della Valle & Livio (1998), who followed up their earlier photometric study by including the spectroscopic classifications of 22 novae with expansion parallax distances, 19 of which were from their 1992 fiducial sample. The major result of their analysis is illustrated in Figure 5.4, where Della Valle and Livio found that the generally faster and brighter He/N novae were concentrated closer to the Galactic plane than were their Fe II counterparts. Very recently, however, Özdönmez et al. (2018) have studied the spatial distribution of a large sample of Galactic novae with known spectroscopic class and found no evidence for the two-nova-population hypothesis. In view of the available evidence, the question of whether the spectroscopic classes of novae vary with distance from the Galactic plane (and hence with stellar population) cannot yet be answered definitively.

5.2 The Galactic Nova Rate

The rate of nova eruptions in the Milky Way is important not only for understanding the formation and evolution of close binary systems, but also to the study of the chemical evolution of the Galaxy. In particular, novae are believed to contribute a significant fraction of certain CNO isotopes such as ^{13}C, ^{15}N, and ^{17}O, as well as being responsible for the production of the radioactive isotopes ^{7}Be, ^{22}Na, and ^{26}Al. Since ^{7}Be beta decays into ^{7}Li with a half-life of about 53 days, novae also contribute to the production of ^{7}Li in the Galaxy. A recent study showed that Nova Sagittarii 2015 No. 2 (V5668 Sgr) produced so much ^{7}Be that two such novae per year would produce enough ^{7}Li to account for all the lithium in the Milky Way (Molero et al. 2016).

The Galactic nova rate has been measured by employing two independent approaches: (1) through an extrapolation of the local nova rate in the solar

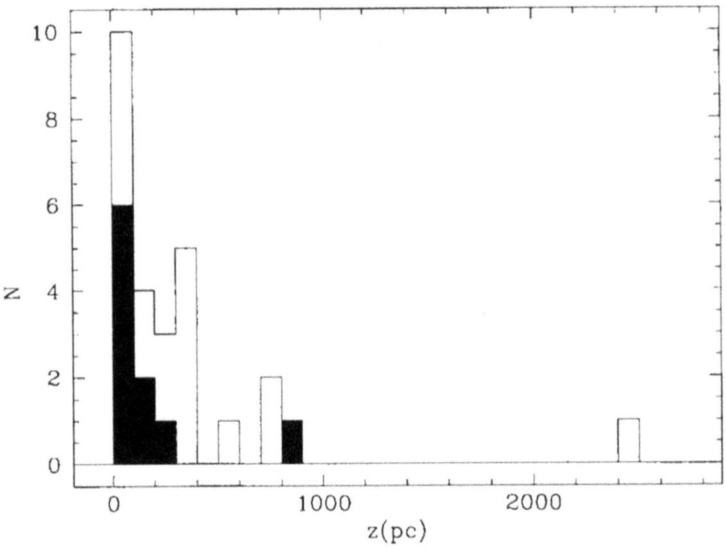

Figure 5.4. The nova frequency distribution as a function of the distance z from the Galactic plane for the sample of novae considered by Della Valle & Livio (1998). The shaded region represents the sub-sample of He/N novae. (Reproduced from Della Valle & Livio 1998. © 1998. The American Astronomical Society. All rights reserved.)

neighborhood to the entire estimated mass of the Galaxy (what we refer to as the "direct" method), and (2) through a suitable scaling of the observed nova rates in extragalactic systems (the "indirect" method).

5.2.1 Direct Determination

A direct estimate of the nova rate in the Galaxy is extremely difficult to obtain given our position situated within the Galaxy. Despite this impediment, as summarized in Table 5.2, there have been several attempts to measure the Galactic nova rate based on an extrapolation of the locally observed rate. One of the pioneering attempts was made by nova-hunter William Liller in collaboration with the accomplished amateur astronomer Benjamin Mayer.[2] Their program, which is described in some detail below, nicely illustrates the challenges faced when trying to measure the Galactic nova rate from an extrapolation of the locally observed rate.

In the early 1980s Liller, who had just retired from Harvard University, and Mayer teamed up to conduct a long-term photographic Galactic nova survey from Liller's new home in Viña del Mar, Chile. The photographs were searched for novae using an instrument Mayer had developed known as the PROBLICOM

[2] Mayer has the distinction of being the first person to have observed (serendipitously) the first Galactic nova (Nova Cyg 1975, V1500 Cyg) as it rose to maximum light. In a stroke of luck, on the night of 1975 August 29, Mayer had been photographing the constellation of Cygnus searching for meteors. He didn't find any and threw the photographs in the trash. Luckily, three days later when he learned that a nova had erupted in Cygnus, the trash hadn't yet been collected and he was able to retrieve the data (Liller et al. 1975).

Table 5.2. Direct Determinations of the Galactic Nova Rate

Reference	Year	Rate (yr^{-1})	Notes
Vorontsov-Vel'yaminov (1948)	1948	200	Extrapolation of Bailey's (1921) estimate of 30 yr^{-1} brighter than $m = 9$
Kukarin (Allen 1954)	1954	100	Extrapolation of local rate
Kopylov (1955)	1955	50	Extrapolation of local rate using M31's radial nova gradient
Sharov (1972)	1972	260	Extrapolating 0.08 yr^{-1} kpc^{-2} at Sun's position using M31's radial nova gradient
Liller & Mayer (1987)	1987	73	Extrapolation of rate in 60° sector of the Galaxy centered on the Sun
Hatano et al. (1997)	1997	41	Monte Carlo simulation using Liller & Mayer rate for novae $m < 11$
Shafter (1997)	1997	35	Extrapolation of local rate in multiple sectors centered on the Sun
Shafter (2002)	2002	30	Extrapolation of local rate $m < 2$ using Bahcall & Soneria model
Shafter (2017)	2017	50	Extrapolation of local rate $m < 2$ using Monte Carlo simulation and updated Galaxy model

(PROjection BLInk COMparator). Prior to Liller's efforts, the discovery of Galactic novae were heavily biased to the northern sky, with many novae being discovered through the singular efforts of famed comet hunter and discoverer of Nova Cyg 1975, Minoru Honda in Japan. Indeed, all but two of the 14 Galactic novae discovered during the period 1978–1982 were found north of declination $\delta = -28°$, six of these by Honda himself. During the first four years of their survey (1983–1986) Liller and Mayer discovered 5 novae south of $\delta = -28°$. During this same period an additional 13 novae were discovered by others, mostly in the northern hemisphere. Thus, during the nine-year period between 1978 and 1986, there were a total of 33 nova candidates discovered, resulting in a raw observed nova rate of 3.67 novae per year. After making corrections for seasonal incompleteness, nova "fade out effect" (essentially the lifetime of nova visibility), incompleteness due to weather, and "overlooking" (essentially human error), Liller and Mayer arrived at a corrected rate of 13.9 novae per year brighter than 11th magnitude.

In order to estimate the Galactic rate, their local nova rate estimate needed to be extrapolated to the entire galaxy. To perform the extrapolation Liller and Mayer first had to estimate distances to the novae in their sample. Since light curve information was not available for the majority of novae, the MMRD relation was of no use, and they were forced to assume that the average nova had an absolute visual magnitude at maximum light of $M_V = -7.7$. Similarly, since extinction estimates for

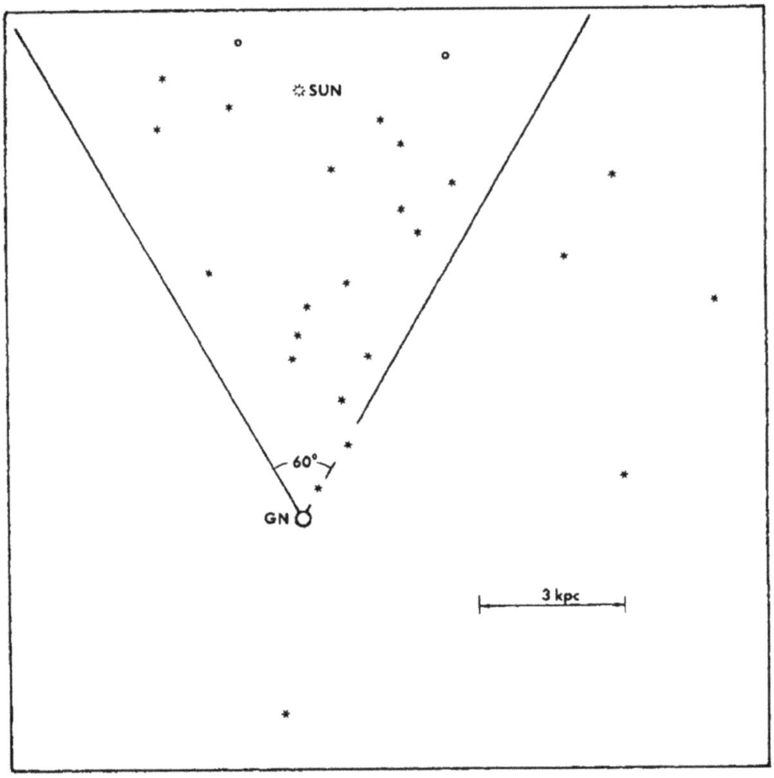

Figure 5.5. The distribution of 25 novae closer than 15 kpc projected onto the plane of the Galaxy. Of these, 19 nova candidates are found within a 60° sector centered on the Sun. Two of the candidates, Cen X-4 and Nova Cep 1963, (open circles), were not considered strong nova candidates and were excluded from the analysis. (From Liller & Mayer 1987. © 1987. The Astronomical Society of the Pacific. All rights reserved. Printed in U.S.A.)

individual novae were lacking, they adopted a mean extinction of $A_V = 0.9$ mag per kpc within 200 pc of the Galactic plane, and $A_V = 0$ everywhere else. Figure 5.5 shows the distribution of the novae considered by Liller & Mayer projected onto the plane of the Galaxy.

Seventeen of the 25 novae estimated to be closer than 15 kpc fell within a 60° sector centered on the position of the Sun. Liller and Mayer reasoned that the Galactic nova rate could be estimated as:

$$n_G = n_{60}\, k_1\, k_2\, k_3 \tag{5.1}$$

where n_{60} is the observed nova rate within the 60° sector, $k_1 = 6$ is the correction factor for the fraction of the Galaxy covered by the sector, k_2 is the correction for extinction within the sector surveyed, and $k_3 = 3.8$ is the correction factor for incompleteness considered above. After estimating $k_2 = 1.7$ based on the number of novae likely missed due to heavy extinction in the direction of the Galactic center behind the Sagittarius arm, Liller and Mayer arrived at the following estimate for the Galactic nova rate:

$$n_G = (17/9) \times 6 \times 3.8 \times 1.7 = 73 \pm 24 \text{ yr}^{-1} \quad (5.2)$$

where the error was estimated from consideration of the uncertainties inherent in the various correction factors. It seems clear from the forgoing discussion that there are tremendous uncertainties inherent in any attempt to extrapolate measurements of the "local" nova rate to the entire galaxy.

A decade after the Liller and Mayer study, Shafter (1997) made a new estimate of the Galactic nova rate using a similar approach. Specifically, he built on the Liller and Mayer study but including a much larger sample of novae with distances determined from the MMRD relation, or, when possible, from nebular expansion parallax. In addition to exploring a range of local nova sector sizes, departures from axial symmetry, as might be expected in the case of a significant galactic bar, were considered. Ultimately, the nova rate determined, 35 ± 11, was roughly half that found by Liller and Mayer.

Another attempt to measure the Galactic nova rate was again attempted by Shafter in 2002. This time, a model of the Milky Way computed by Bahcall & Soneria (1980) was used to extrapolate the local nova rate to the entire Galaxy. Number counts of stars as a function of apparent magnitudes from the Bahcall and Soneria model were compared with the cumulative nova rate as a function of magnitude. Locally (for novae with $m < 2$), the nova rate and stellar density both increased with increasing apparent magnitude as expected for a disk distribution (i.e., $N(m) \propto 10^{0.4m}$). This result led Shafter to assume that the observed nova rate was essentially complete for novae brighter than $m = 2$, and resulted in an overall Galactic nova rate of ~ 30 yr^{-1}. As pointed out years later by Bradley Schaefer, who is especially knowledgeable in the history of Galactic nova observations, the historical record of discovery of novae brighter than second magnitude is likely incomplete, with as many as half of these novae being missed. If so, then the Galactic nova rate could possibly be as high as 60 per year.

A recent attempt to estimate the Galactic nova rate directly was undertaken by Shafter in 2017 (Shafter 2017). The rate was estimated using two approaches: the first being essentially a re-implementation of the 2002 analysis based on a larger sample of novae, and the second introduced a Monte Carlo numerical simulation approach to estimate the Galactic nova rate required to produce the observed number of novae brighter than second magnitude. A major advantage of the Monte Carlo approach is that it allowed the uncertainties in the observed number of novae and in the assumed completeness at bright magnitudes to be propagated to the final Galactic nova rate estimate. Figure 5.6 shows the result of the Monte Carlo analysis that yielded a Galactic rate of 50^{+31}_{-23} yr^{-1}.[3]

[3] Özdönmez et al. (2018) described following a prescription similar to that employed by Shafter a year earlier, and found a slightly higher Galactic nova rate of 67^{+21}_{-17} yr^{-1}.

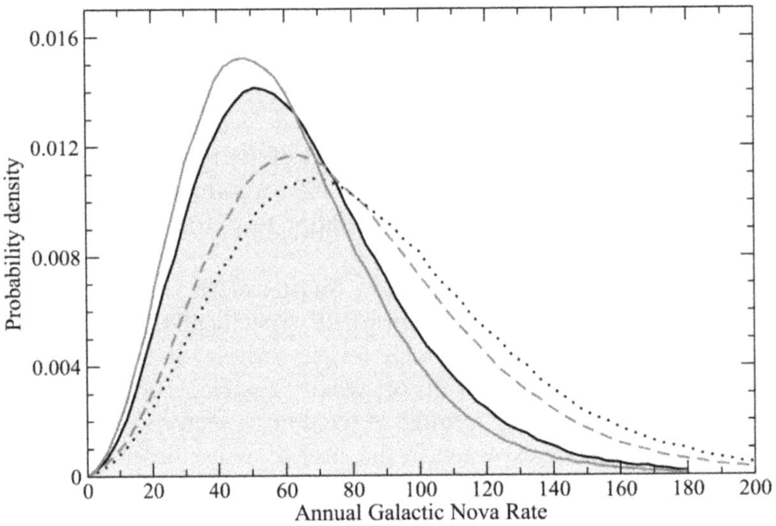

Figure 5.6. Galactic nova rate probability density distributions from Monte Carlo simulations. The shaded region bounded by the black solid line represents the preferred model, which yields a best estimate Galactic nova rate of 50^{+31}_{-23} yr^{-1} (shaded region). The red curve, dashed and dotted lines represent different (less likely) model assumptions (see Shafter 2017 for details). (Reproduced from Shafter 2017. © 2017. The American Astronomical Society. All rights reserved.)

The rather large uncertainties seen in Figure 5.6 confirms the difficulty involved estimating the Galactic nova rate from an extrapolation of the local rate, which is based on a small number of relatively nearby novae.

5.2.2 Scaling from External Galaxies

The other, and arguably more promising, approach to measuring the Galactic nova rate is through scaling the luminosity-specific nova rate (LSNR) measured in other galaxies to the luminosity of the Milky Way. Although application of the method seems straightforward, there are a number of difficulties with its implementation. Not only does it require a reliable estimate for the LSNR typical of galaxies similar to the Milky Way (i.e., Sbc spirals), it also requires a reliable estimate for the K-band luminosity of the Milky Way. At the present time, both quantities are quite uncertain. Nevertheless, throwing caution to the wind and adopting an absolute visual magnitude for the Galaxy $M_V \sim -21$, a V–K color of ~3.2 typical of Sb galaxies (Grasdalen 1975), and an absolute K-band magnitude for the Sun of 3.3 (Pecaut & Mamajek 2013), we find an estimated Galactic K-band luminosity of $\sim 10^{11}$ $L_{K,\odot}$. Adopting an average LSNR of $\nu_K = 2.3$ yr^{-1} $[10^{10}$ $L_{K,\odot}]^{-1}$ Shafter et al. (2014) yields a Galactic nova rate of ~ 23 per year. A comparison of K-band luminosity of the Milky Way with the $LSNR_K$ for M31 alone yields a slightly higher nova rate of ~30 novae per year. This value is about 60% of the latest value estimated from Galactic data (Shafter 2017). A likely explanation for the discrepancy between the direct and indirect Galactic nova rate estimates will be discussed in the following two chapters. In particular, recent evidence is emerging that the average LSNR for galaxies with measured nova rates may be underestimated by a factor of 2–3.

References

Allen, C. W. 1954, MNRAS, 114, 387
Bahcall, J. N., & Soneira, R. M. 1980, ApJS, 44, 73
Buscombe, W., & de Vaucouleurs, G. 1955, Obs, 75, 170
Ciardullo, R., Ford, H. C., Neill, J. D., et al. 1987, ApJ, 318, 520
Darnley, M. J., Bode, M. F., Kerins, E., et al. 2006, MNRAS, 369, 257
Della Valle, M., Bianchini, A., Livio, M., & Orio, M. 1992, A&A, 266, 232
Della Valle, M., & Livio, M. 1998, ApJ, 506, 818
Duerbeck, H. W. 1987, SSRv, 45, 1
Duerbeck, H. W. 1990, in IAU Coll. 122, Physics of Classical Novae, ed. A. Cassatella, & R. Viotti (Berlin: Springer), 34
Eyres, S. P. S., Evans, A., Zijlstra, A., et al. 2018, MNRAS, 481, 493
Grasdalen, G. L. 1975, ApJ, 195, 605
Hatano, K., Branch, D., Fisher, A., & Starrfield, S. 1997, ApJ, 487, 45
Kamiński, T., Menten, K. M., Tylenda, R., et al. 2015, Natur, 520, 322
Kopylov, I. M. 1955, IzKry, 13, 23
Liller, W., Shao, C. Y., Mayer, B., et al. 1975, IAU Circ. #2848
Liller, W., & Mayer, B. 1987, PASP, 99, 606
Molero, P., Izzo, L., Bonifacio, P., & Della Valle, M. 2016, MNRAS, 463, L117
Özdönmez, A., Ege, E., Güver, T., & Ak, T. 2018, MNRAS, 476, 4162
Pecaut, M. J., & Mamajek, E. E. 2013, ApJS, 208, 9
Shafter, A. W. 1997, ApJ, 487, 226
Shafter, A. W., & Irby, B. K. 2001, ApJ, 563, 749
Shafter, A. W. 2002, AIPC, 637, 462
Shafter, A. W., Curtin, C., Pritchet, C. J., et al. 2014, in ASP Conf. Ser. 490, Stella Novae: Past and Future Decades, ed. P. A. Woudt, & V. A. R. M. Ribeiro (San Francisco, CA: ASP), 77
Shafter, A. W. 2017, ApJ, 834, 196
Sharov, A. S. 1972, SvA, 16, 41
Vorontsov-Velyaminov, B. A. 1948, Gaseous Nebula and Novae (Moscow: Leningrad, Izd. Akad. Nauka), (in Russian)

AAS | IOP Astronomy

Extragalactic Novae
A historical perspective
Allen W Shafter

Chapter 6

Modern Nova Surveys of M31

By the early 1980s, when the photographic surveys of Rosino, and of Sharov and Alksnis, were winding down, four bright emission-line sources were discovered serendipitously in Hα image tube plates taken as part of a search for planetary nebulae near the nucleus of M31 (Ciardullo et al. 1981, 1983; Ford et al. 1981, 1982). A reproduction of the Hα image tube plate is shown in Figure 6.1, with the positions of the four Hα sources marked. Initially, the nature of the these sources was unclear, but subsequent spectroscopy with the 3 m Shane reflector at Lick observatory established that they were indeed novae shortly after maximum light (see Figure 6.2).

The discovery of four M31 novae on a single image tube plate suggested that Hα imaging surveys might be a very efficient technique for the discovery of extragalactic novae. This realization, coupled with the proliferation of CCD imaging cameras, led to a renaissance in the study of extragalactic novae. The digital format, high quantum efficiency, and linear response of the new CCD detectors made it possible for the first time to do precise photometry of faint novae against the bright background of M31's bulge. In addition, the enhanced red response of CCD detectors relative to the most sensitive of the photographic plates made Hα imaging particularly efficient.

Despite their obvious advantages, the early CCDs were not ideal imaging devices, with the principal drawback being their small format and resulting limited areal coverage. In the early 1980s the workhorse was the 320 × 512 pixel RCA (SID 53601) chip with 30μ pixels. At a minimal spatial resolution of 0.4″ per pixel necessary to resolve stars under good seeing conditions, these devices provided areal coverage of just 2.1′ × 3.4′. Thus, to cover a good portion of M31's bulge (~10′ × 10′) required at least a dozen individual fields. These early limitations aside, the characteristics of the CCDs steadily improved, and quickly replaced photographic plates as the dominant detector in optical astronomy.

doi:10.1088/2514-3433/ab2c63ch6

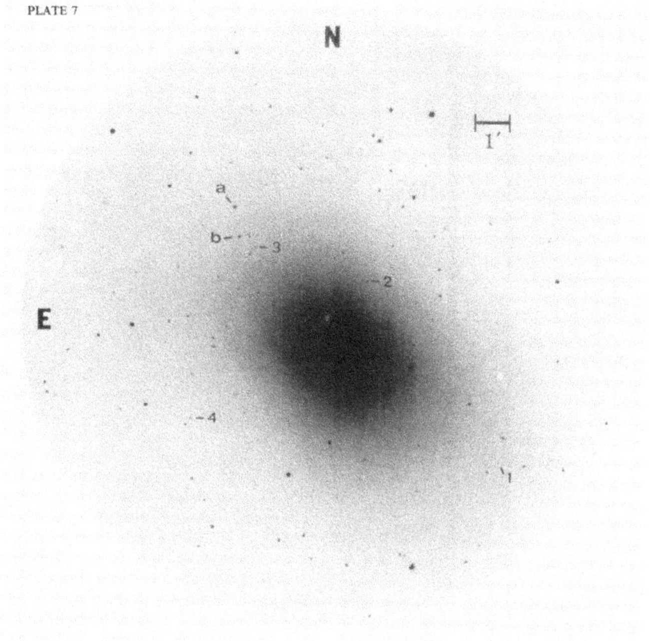

Figure 6.1. A reproduction of the image tube plate from the Ciardullo et al. (1983) M31 planetary nebula survey showing the discovery of four Hα sources later identified as novae. Stars **a** and **b** are secondary standards. (Reproduced from Ciardullo et al. 1983. © 1983. The American Astronomical Society. All rights reserved.)

6.1 Hα Nova Surveys of M31

6.1.1 The Ciardullo et al. Survey

In a pioneering effort, Ciardullo et al. (1987) undertook the first systematic Hα CCD survey for novae in M31 over a period of ~5 years using the Kitt Peak National Observatory (KPNO) #1 0.9 m telescope, the McDonald Observatory's 0.8 m telescope, and the 1.0 m reflector at the University of Tel Aviv's Wise Observatory. In their first effort, Ciardullo et al. analyzed the subset of the data taken at KPNO, which covered a 26.6′ by 15.6′ region along the major and minor axis of M31.

A total of 35 nova candidates were discovered that led to two significant insights into the nova population in M31. First, the increased contrast that the Hα images provided against the background light of the galaxy allowed the detection of novae all the way to within 10″ of the center of the galaxy. This discovery refuted the assertions of Hubble (1929), and later Arp (1956) and Rosino (1964) that a "nova hole" existed near the nucleus of M31. The second conclusion of the Ciardullo et al. study was that the spatial distribution of a combined sample of M31 novae that included the Hubble and Arp samples followed the bulge light more closely than the overall light of the galaxy (see Figure 6.3). Indeed, a comparison of the data with

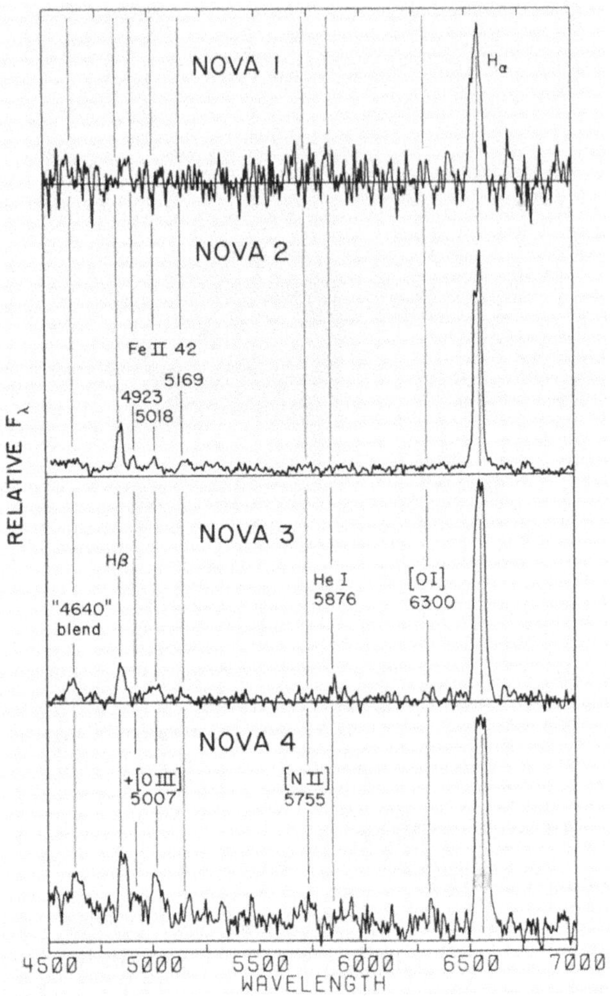

Figure 6.2. Spectra of the four Hα sources discovered by Ciardullo et al. (1983) on the plate shown in Figure 6.1. Note the extremely strong Hα emission lines that develop shortly after maximum light. Imaging in Hα aids in the discovery of novae by providing enhanced contrast against the bright galaxy background. (Reproduced from Ciardullo et al. 1983. © 1983. The American Astronomical Society. All rights reserved.)

bulge and disk models of M31 showed that the density of novae in the bulge exceeds that in the disk by at least an order of magnitude, and that the best fit was achieved when *all* novae were assumed to arise from the bulge. This surprising result led Ciardullo et al. to speculate that the number of nova progenitor binaries in the bulge had been enhanced over the field by a contribution from nova systems spawned in the dense cores of M31's globular clusters. It was thought that these nova progenitor binaries could subsequently be ejected into the bulge via three-body encounters within the clusters, or through tidal disruption of the clusters, or both.

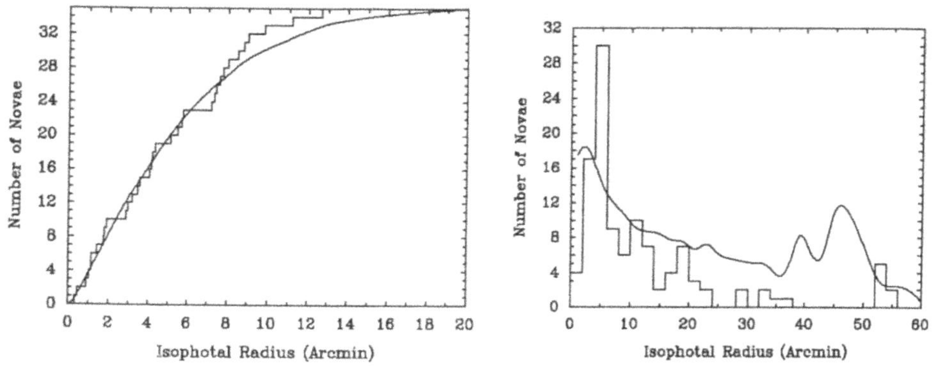

Figure 6.3. (Left panel) The cumulative distribution of the Ciardullo et al. (1987) nova sample compared with the cumulative *B* light of M31. The cumulative nova distribution follows the background light extremely well at small radii, but falls off faster than the background light at large radii. (Right panel) Comparison of the radial nova distribution of the Hubble–Arp sample of novae with the background *B*-band light. The nova distribution is significantly more centrally concentrated than is the background light. (Reproduced from Ciardullo et al. 1987. © 1987. The American Astronomical Society. All rights reserved.)

6.1.2 Exploring the Disk of M31

Following the success of the Ciardullo et al. survey of the bulge region of M31, starting in 1990 Shafter & Irby (2001) undertook two long-term surveys of novae in M31 using the 40 inch reflector at San Diego State University's Mount Laguna Observatory. These studies were novel in that they were the first designed specifically to explore possible population differences by measuring the nova rate in the disk and bulge of M31 separately. The first survey (Survey A) took place between 1990 September and 1993 December, and covered a total of 11 separate fields in the bulge of M31 extending ~30′ either side of the nucleus along the major axis. The second survey (survey B), conducted between 1995 August and 1997 November, covered five fields, one centered on the nucleus of M31, with the other four situated along the major axis approximately one degree from the nucleus (two to the NE and two to the SW), well out into the disk of the galaxy. The positions of the survey fields are shown in Figure 6.4.

Shafter & Irby discovered a total of 82 novae during the course of their surveys, 53 and 29 in surveys A and B, respectively. Following Ciardullo et al. (1987), Shafter & Irby compared the observed spatial distributions of the novae with the overall background light, and also with estimates of the bulge and disk light separately. In the case of both surveys, a fit of the spatial distribution of the nova samples confirmed the earlier conclusion reached by Ciardullo et al. (1987) that the novae were primarily associated with the bulge of M31. To quantify the results, a maximum likelihood analysis was performed, showing that the best fit was achieved if the bulge produced novae at ~2.5 times the rate per unit light compared with the disk ($\theta = 0.4$). After correcting for both spatial and temporal incompleteness in their surveys, Shafter & Irby (2001) determined a global nova rate for M31 of 37^{+12}_{-8} per year, which is ~25% higher than that found in earlier investigations. Under the

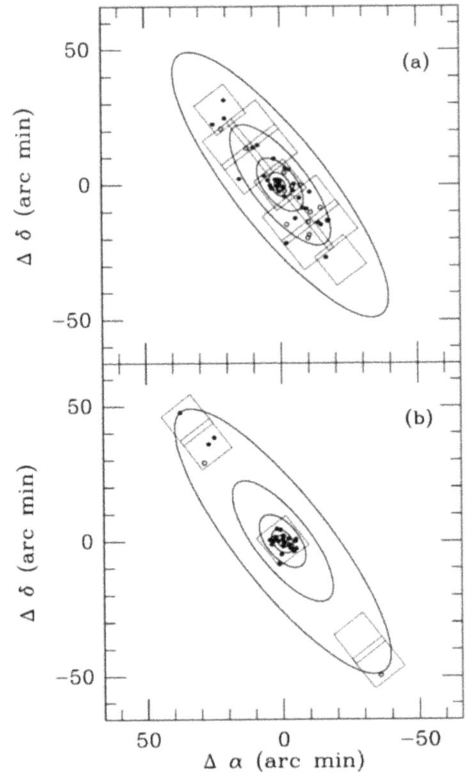

Figure 6.4. The positions of the 82 novae discovered in the Shafter & Irby survey (2001). Panel (a) shows the positions of 53 novae in 11 separate fields that composed the first part (1990–1993) of the survey, and panel (b) shows the positions of 29 novae from the second (1995–1997) survey. (Reproduced from Shafter & Irby 2001. © 2001. The American Astronomical Society. All rights reserved)

assumption that the disk produced novae as efficiently as the bulge ($\theta \simeq 1$), the M31 nova rate could approach 50 yr^{-1}.

A potential complication in the comparison of the specific bulge-disk nova rates is that extinction internal to M31 may suppress the observed disk nova rate. As a check on this possibility Shafter & Irby (2001) compared the observed spatial distribution of planetary nebulae with that of the novae in M31. Unlike the novae, which appear to follow the bulge light, the observed planetary nebula spatial distribution followed the overall (bulge + disk) background light of the galaxy remarkably well. Since the novae were discovered from their Hα emission, while the planetary nebula were discovered based on their shorter-wavelength [O III] λ5007 Å emission, if extinction was playing a major role in modifying the observed spatial distribution it should affect the planetary nebulae more than the novae. Thus, a robust conclusion is that the actual nova distribution is less spatially extended than the planetary nebula distribution. Since the PNe birthrate is not expected to be sensitive to stellar population, the association of novae with the bulge of M31 seems to be real, and not a result of extinction in M31's disk.

6.2 The Role of Amateur Astronomers

Amateur astronomers have enjoyed a long history in the discovery of astronomical transients. Comet hunters who image dawn and dusk sky in search of their targets have for decades serendipitously discovered numerous Galactic novae. At the other end of the cosmic distance scale, Robert Owen Evans, minister of the Uniting Church of Australia and amateur astronomer, is renowned for his discovery of extragalactic supernovae. What is particularly remarkable is that Evans discovered the supernovae by searching for new "stars" in the fields of some 1500 galaxies that he had committed to memory! At the time of writing, Evans still holds the record for the number of visual discoveries of supernovae, 42 in total, during the period from 1981 through 2008.

It is difficult to overemphasize the important role that amateur astronomers have played and are increasing playing in the study of novae in M31. An early pioneer in the discovery of nova in M31 was James Bryan, an avid amateur astronomer from Georgetown, Texas, who began imaging M31 in 1987. Equipped with only a 35 mm SLR camera, Bryan discovered a number of novae in M31, including one object (M31N 1990-10a) that turned out to be a likely recurrent nova, one of only about 20 such objects so far identified in M31 (Shafter et al. 2015).

Despite the contributions of amateurs such as Bryan with photographic imaging, as is the case of professional astronomy, progress is heavily influenced by the advance of technology. In particular, the increasing availability of CCD imagers and computerized telescopes in the 0.4–0.6 m range made the discovery of novae in local group galaxies, particularly M31, quite feasible for serious amateur astronomers.

One such astronomer is Kamil Hornoch who began his amateur astronomy career in 1985 observing a variety of transient sources including comets, asteroids, and variable stars from his home village of Lelekovice (pop ~1400) in the Czech Republic.[1] Hornoch started out armed with only a pair of 10×50 binoculars, but within a year he was able to upgrade to a 5 inch reflector, which he used for the next eight years. In 1994 he acquired a much more powerful 14 inch (35 cm) reflector, allowing him to extend his observations to fainter transients. Since 2002, Hornoch has been focusing his efforts on the discovery of novae in M31, where he has discovered 185 novae through the end of 2017, or almost one out of five of all novae ever discovered in that galaxy—more than any single person in history!

6.2.1 The RBSE Survey

In the mid 1990s an NSF-funded educational program to expose secondary school teachers to astronomical research was initiated by Suzanne Jacoby, Donald McCarthy, and Jeffrey Lockwood at KPNO. It was decided that an ideal project for the resulting Research-Based Science Education (RBSE) program would be to use the KPNO #2 0.9 m telescope to search for novae in M31. Following the success of the Ciardullo et al. (1987) survey, the RBSE effort employed Hα imaging to

[1] Hornoch has been employed as a professional astronomer at the Ondřejov Observatory in the Czech Republic since 2007.

conduct their survey. Although some preliminary observations were obtained as early as 1995, the program formally commenced in 1998, with initial funding adequate to continue the program for three years. This pilot program proved so successful that the RBSE program was continued into the new millennium. In 2001 Travis Rector took charge of the project and has run the program for the past 20 years, during which time more than 200 epochs of M31 observations have been covered. Within a few years, the 0.9 m telescope became associated with the Wisconsin, Yale, Indiana (WYIN) telescope consortium, at which point the focus of the program switched to mentoring undergraduate students. A total of 287 novae have been detected in images taken thus far during the course of the project, with 71 of them discovered by the RBSE project.

One of the drawbacks of the RBSE project was that funding for the program did not provide support for thoroughly analyzing and publishing results from the survey. As a consequence, little in the way of a systematic analysis of the data has yet appeared in print, which is a shame given the wealth of data accumulated over the more than two decades that the program has been active. Fortunately, the situation is improving, with preliminary analyses now starting to appear in the literature (Crayton et al. 2018). The RBSE data set, with its broad spatial and temporal coverage, is ideal for studying the spatial distribution of novae in M31.

6.3 Microlensing Surveys

Starting in the 1990s a major effort was initiated to study microlensing events in the halos of both the Galaxy and M31. Microlensing results from the gravitational lensing and subsequent magnification of a background light source (usually a star from a high surface brightness region like the bulge of the Milky Way or of a nearby galaxy like M31) by a relatively faint foreground compact object. These surveys, which involve wide-field imaging at high cadence, can be exploited for the discovery of unlensed, but variable sources such as novae.

One such project, POINT-AGAPE (Pixel-lensing Observations with the Issac Newton Telescope—Andromeda Galaxy Amplified Pixel Experiment) was used by Liverpool John Moores University doctoral student Matthew Darnley and collaborators (Darnley et al. 2004) to discover a total of 20 nova candidates in M31 from a survey period between 1999 and 2002. The data from this survey provided uniform coverage and depth of two 0.3 deg^2 fields to the immediate north-east and south-west of M31's nucleus, and thus provided a good sample to study the spatial distribution and rate of novae in the bulge of the galaxy. In contrast to previous surveys where novae had been detected primarily through a visual examination and comparison of the individual images (a process known as "blinking"), Darnley et al. (2004) used an automated data reduction procedure where they performed aperture photometry on more than a quarter million point sources from the POINTE-AGAPE data set. They then produced light curves for transient sources displaying variability consistent with that expected for a nova. After conducting a thorough test for incompleteness in their detection pipeline, Darnley et al. (2006) computed a global nova rate of 65^{+16}_{-15} novae per year in M31. This value, which is approximately 50% higher than previous estimates, is still widely regarded as the best estimate of the nova rate in M31 currently available.

6.4 The Automated Surveys

Just prior to the dawn of the new millennium an ambitious new project was proposed to study the transient sky—The Large Synoptic Survey Telescope (LSST). The LSST was envisioned to be a large aperture, wide-field telescope that would be dedicated to the survey the entire night sky as viewed from Cerro Pachón, Chile approximately twice per week to an unprecedented depth of 24th magnitude. The astronomical community quickly got behind the LSST project, and the 2010 decadal survey put the LSST as the top priority for large-scale projects in astronomy.[2]

A metric, entendue—the product of the areal coverage (deg^2) and the aperture (m^2)—was adopted to describe the "power" of such surveys. The entendue of the 8.4 m LSST would be ~320 m^2 deg^2, much larger than any competing survey currently operational. The primary science driver for the LSST was to discover a variety of transient sources including, but not limited to, near-Earth asteroids, eclipsing binary stars, variable stars, novae and supernovae.

Since its inception, there have been several challenges faced by the LSST project. Among these were: (1) the survey would not have access to sky north of $\delta \sim 30$ deg, (2) the enormous amount of data collected by the LSST on a typical night of observing was estimated to be of order 3 terabytes, and the data storage and computing resources required had never been accommodated before, and (3) the scale and associated cost of the project was such that it was estimated to take ~15 years from ground-breaking to completion.

Given these challenges, by the period roughly between 2005 and 2010 several complementary and competing projects were proposed as forerunners to the LSST. Two studies in particular, which have particular relevance to the study of novae in M31, are described below.

6.4.1 Pan-STARRS

The Panoramic Survey Telescope and Rapid Response System (Pan-STARRS) consists of two 1.8 m telescopes (PS1 and PS2) located near the summit of Haleakala on the Island of Maui. The project, which is led by the University of Hawaii in collaboration with 13 other research institutions in six countries and funded by the United States Air Force, has a primary mission of detecting near-Earth objects (NEOs) such as asteroids that could pose a hazard to Earth. The PS1 telescope started science operations in 2010, with PS2 becoming operational in 2017. The wide field camera used on both telescopes has approximately 1.4 billion pixels, and can image approximately 6000 square degrees per night with typical exposure times of about a minute. After the initial three years of operation, the PS1 telescope had imaged the entire sky visible from Haleakala a dozen times in five separate filters (g,r,i,z,y). Although not fully exploited for nova studies as yet, the wealth of data

[2] http://www.nap.edu/catalog/12951/new-worlds-new-horizons-in-astronomy-and-astrophysics

from the Pan-STARRS project promises to improve our understanding of the properties of novae both in the Galaxy and beyond.

6.4.2 PTF and its Successors

The Palomar Transient Factory (PTF) project was the brain-child of Shrinivas Kulkarni (circa 2007), shortly after he became director of the Caltech Optical Observatories. Kulkarni, recognized early that given recent advances in detector technology and computer storage capabilities, time-domain surveys were poised to play a prominent role in the study of the dynamic sky, and that much could be done prior to the completion of the ambitious LSST project. A key to the project was the availability of the venerable Samuel Oschin 48 inch Schmidt telescope, the same telescope used to conduct the legendary Palomar Sky Survey back in the 1950s. The wide field of view of the Schmidt design made the 48 inch Oschin telescope well suited to the task. The telescope was retooled for robotic operation and a modern 7.26 deg^2 mosaic CCD camera was installed (Law et al. 2009). Follow-up observations were conducted using another telescope already available on Palomar mountain, the fully automated 60 inch reflector. The scientific scope of the PTF project was broad and open-ended—to explore the variable night sky to a depth of $m_R \simeq 21$ with cadences ranging from the order of minutes for some fields to years in the case of others (Rau et al. 2009).

The PTF program officially began operation in 2009 March, and continued until the end of calendar year 2012. During this period, a number of important discoveries were made including the discovery of supernova 2001fe sooner after eruption that any Type Ia supernova had ever been observed. A broader goal of the survey was to systematically explore the range of transients in the luminosity–timescale plane. This work, which formed the core of the doctoral thesis of Mansi Kasliwal, led to the characterization of a number of previously unknown classes of transients such as the Ca-rich transients or "gap" transients whose luminosities lie between nova and supernovae (see Figure 6.5).

The successor project to the PTF, the intermediate Palomar Transient Factory (iPTF), which included upgrades to the data reduction and source identification software, was implemented at the beginning of 2013. During these years, the telescopes were in operation nearly every clear night, with M31 being a frequent target. Due to the vagaries of the weather and scheduling considerations, continuous nightly coverage of M31 was not possible. Nevertheless, the resulting data provide one of the best opportunities for the study of novae in M31 currently available.

Until recently, very few nova studies from the PTF or iPTF have appeared in print. Soraisam et al. (2017), analyzed a portion of the iPTF data obtained in the fall of 2013 observing season (2013 September–2014 January), as part of a broader effort to develop an automated search algorithm for the detection of transient sources in high cadence optical surveys. During the five-month span of data analyzed, they detected a total of 15 nova candidates, 12 of which had been discovered previously. Of the three potential new novae discovered, one appears to be clearly a nova. The

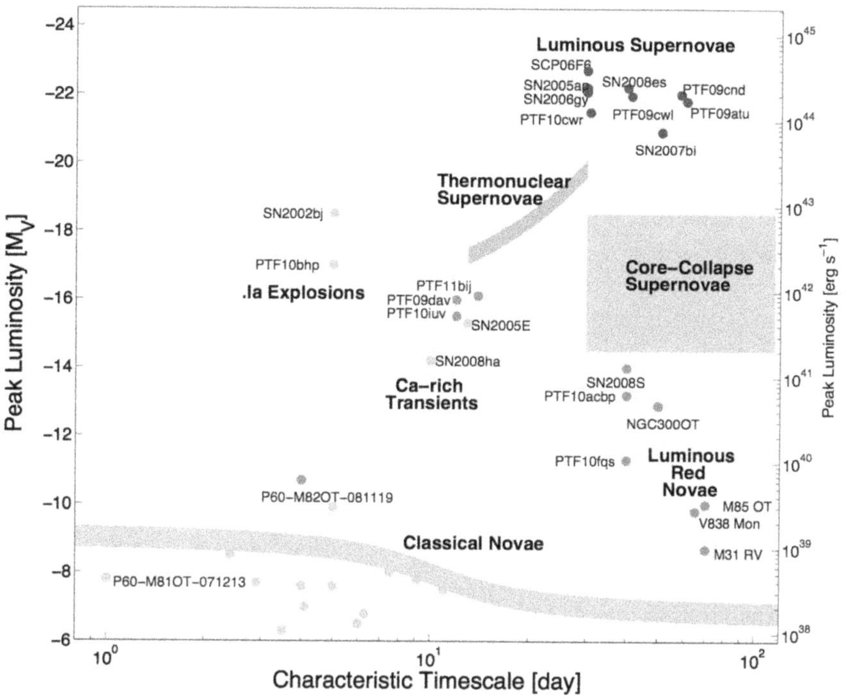

Figure 6.5. Transients in the luminosity–timescale plane (Kasliwal 2011, 2012). Note the diversity of transients populating the parameter space. (With permission from Caltech Institute of Technology.)

other two objects are candidate members of the "faint and fast" group, but the detections happened near the end of the data set, and their nova natures are uncertain.

There remains a wealth of data in the PTF and iPTF archives that is waiting to be thoroughly exploited. In addition, the latest incarnation of the survey boasting a new camera with an amazing 47 deg^2 field-of-view, the Zwicky Transient Facility (ZTF) just initiated routine operation in 2018. The ZTF survey is expected to make major contributions to the study of extragalactic novae.

6.5 Spectroscopic Surveys of Local Group Novae

The extragalactic nova studies described thus far have all been imaging surveys—first photographic and later digital (CCD) studies. Despite the wealth of photometric data, spectroscopic observations of extragalactic novae have only recently become widely available. In the early years of extragalactic nova research there were several practical and structural impediments making it difficult to acquire spectroscopic observations of extragalactic novae. At the distance of M31, novae are relatively faint even near maximum light ($m > 15$), and their eruptions are, of course, unpredictable. Thus, follow-up spectroscopic observations required large telescopes (>3 m aperture for moderate resolution spectra), that needed to be scheduled quickly. These requirements run against the grain of traditional telescope scheduling protocols, where telescope time was typically allocated in relatively short runs (e.g., 3–5 nights), and was done months in advance.

Despite the scheduling challenges, it was clear that spectroscopic observations had the potential to greatly extend our understanding of novae in different stellar populations. The first description of spectra of novae in M31 dates back to 1932 in a paper by Humason (1932). In this early paper, Humason describes in some detail the spectra of two bright novae (novae #106 and #108) that were both discovered by Hubble prior to maximum light (m_{pg}[max] = 15.5 and 15.8, respectively). An extrapolation of the lightcurves given by Humason suggests that both novae were relatively fast, with t_2 times of order 2 weeks. The first nova, now designated M31N 1932-07a, was characterized by relatively broad Balmer emission lines with P Cygni absorption features on the blue wings. The mean widths of the Hβ and Hγ lines (54 Å) correspond to velocity widths of $\sim$3500 km s^{-1}. The second nova, now designated M31N 1932-09e, was discovered by Hubble on 1932 September 24, and reached maximum light approximately two days later. The Balmer emission lines were narrower than in 1932-07a, and were characterized by a full width at half maximum (FWHM) of $\sim$1400 km s^{-1}. In the classification scheme of Williams (1992) described in Chapter 4, M31N 1932-07a would likely be classified as a broad-lined Fe II nova, while 1932-09e appears to be a typical Fe II nova. In his paper, Humason mentions that he had obtained an earlier slit-less spectrum of the brightest nova in Hubble's original survey, Nova #54, which reached an apparent magnitude of m_{pg} = 15.3 on 1925 September 18, approximately six days after discovery. The spectrum, which revealed bright (emission) lines at Hα and Hβ is believed to be the first spectrum of an M31 nova ever taken.

After these pioneering spectroscopic observations, there were no spectra of novae in M31 published for almost half a century! In the fall of 1978, Dopita et al. (1979) were conducting a photographic survey for supernova remnants in the disk of M31 using the Palomar 48 inch Schmidt and reported the discovery of a likely nova on the night of 1978 August 27–28. The object, which reached $B = 16.3$ at the beginning of September based on plates taken at Asiago Observatory, faded quickly in B but remained bright in Hα for at least a week. A spectrum, obtained on the night of 1978 September 9 with the Palomar 200 in, showed relatively narrow emission Balmer (Hα and Hβ) lines along with very weak emission features near 4160 Å, 4722 Å, and 5082 Å. Although, the spectrum was not published, and the details are somewhat sketchy, the observations appear consistent with the object, now identified as M31N 1978-08a, being a fast nova in M31.

The first spectrum of a nova from the Ciardullo et al. survey was published by Cowley & Starrfield (1987). The nova, #32 in the survey (later dubbed M31N 1986-08a) turned out to be a quite unusual object. It was an extremely slowly-evolving nova ($t_2 \sim 3$ yr), and a member of the Fe II spectroscopic class (see Figure 6.6). The object faded so slowly that it was still visible four years after discovery (see the long-term light curve in Figure 6.7). The nova was studied extensively by Tomaney & Shafter (1993), who concluded that it ejected of order 5×10^{-5} M$_\odot$ of material with nearly solar abundances.

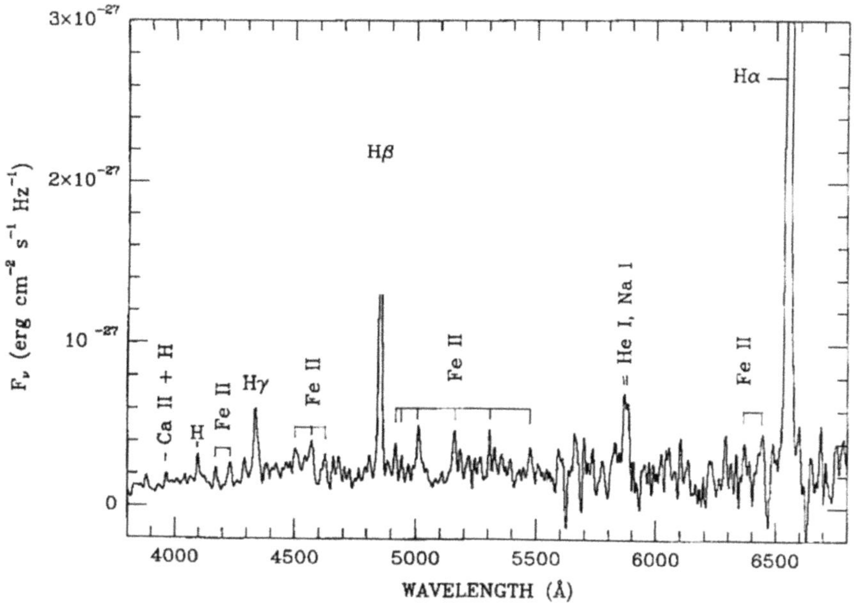

Figure 6.6. A spectrum of M31N 1986-08a. The spectrum is characterized by particularly strong and narrow Balmer and Fe II emission features, and is a typical example of Fe II class nova. The spectrum has the distinction of being the first ever published of a nova in M31. (From Cowley & Starrfield 1987. © 1987. The Astronomical Society of the Pacific. All rights reserved. Printed in U.S.A.)

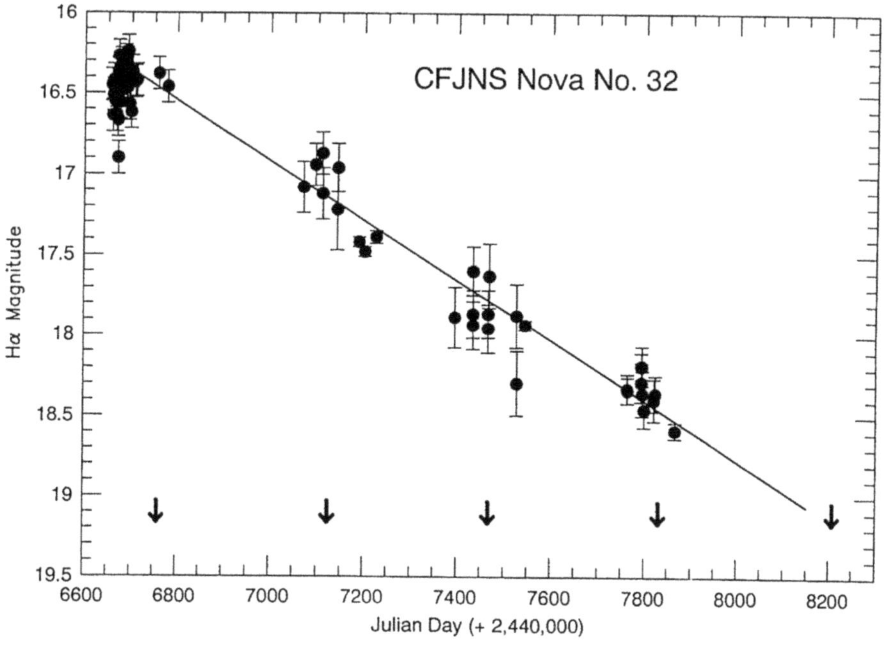

Figure 6.7. The light curve of M31N 1986-08a. The photometric evolution of the nova is one of the slowest on record with $t_2 \sim 3$ yr. The arrows indicated time when spectroscopic observations were acquired. (Reproduced from Tomaney & Shafter 1992. © 1992. The American Astronomical Society. All rights reserved.)

As described earlier, the mid 1980s saw a resurgence in the study of novae in M31. In addition to the Ciardullo et al. study, automated supernova searches, such as the pioneering Katzman Automated Imaging Telescope (KAIT) at Lick Observatory, were getting underway by the start of the next decade. The increased monitoring of M31 with the KAIT, in particular, led to the discovery of several novae in the early 1990s. These surveys, along with the increased availability of low-resolution CCD spectrographs made it practical for the first time to do follow-up spectroscopy of many of the novae gleaned in these surveys.

The first spectroscopic survey was begun at the University of Texas' McDonald Observatory in the fall of 1987. The project was led by Austin B. Tomaney as part of his doctoral thesis under the guidance of E. L Robinson and A. W. Shafter. Targets for the spectroscopic survey were also discovered at McDonald as part of a companion imaging survey using the Observatory's 0.8 m reflector. The survey continued for the next two years with a total of 23 low-resolution (10 Å) spectra of nine novae obtained using the 2.7 m reflector coupled with the Large Cassegrain Spectrograph (Tomaney & Shafter 1992). Spectroscopic observations of the novae were obtained at varying times post maximum light, ranging from as little as a few days to as long as several years post eruption, as in the case of the extremely slow nova M31N 1986-08a (Tomaney & Shafter 1993). Of the nine novae with spectra, it was possible to determine spectroscopic types for eight of them—seven Fe II novae and one He/N nova.

Spectroscopic observations of novae in M31 continued sporadically over the next decade or so, and then in 2006, Shafter and collaborators initiated a major spectroscopic survey of novae in M31 using the 9.2 m Hobby–Eberly Telescope (HET) at McDonald Observatory. We will return to this topic in our discussion of nova populations in Chapter 7.

6.6 X-ray and Infrared Surveys of Novae in M31

Although the subject of this book is focused primarily on optical observations of novae, X-ray observations, and to a lesser extent infrared observations, have played an important role in our understanding of novae. Thus, before leaving our discussion of modern nova surveys in M31, a brief review of these topics is in order.

6.6.1 X-ray Emission in Novae

Novae produce X-ray emission through three basic mechanisms: (1) accretion of matter onto the white dwarf during the quiescent phase, (2) shocked plasma emission during the eruption as the nova ejecta encounter circumstellar gas or a wind from the secondary star, and (3) thermal soft X-ray emission from residual nuclear burning on the surface of the white dwarf. A comprehensive discussion of X-ray emission from accreting white dwarfs can be found in the recent review by Koji Mukai (2017).

In the first case, roughly half of the available accretion energy is radiated through the accretion disk that typically spans a temperature range of several times 10^3–10^4 K. Thus, the dominant disk radiation emerges at infrared, optical and ultraviolet wavelengths. The other half of the accretion energy is available at the boundary layer where the Keplerian accretion disk flow impacts the more slowly rotating white dwarf. Significantly higher temperatures are expected to be reached in the boundary layer, leading to X-ray emission. If the accretion rate is sufficiently low, the boundary layer and environs will be optically thin, and the gas will be unable to cool efficiently. This situation leads to temperatures of $\backsim 10^8$ K and the production of hard X-rays ($E \gtrsim 10$ keV). Higher accretion rates and more optically thick conditions in the boundary layer produce softer X-rays and extreme UV radiation characteristic of black-body temperatures near $\backsim 10^5$ K (Patterson & Raymond 1985a, 1985b; Orio et al. 2001).

In the second case, hard X-ray emission can be produced as a result of the strong shock that results when the nova ejecta encounters ambient gas in the system. This gas can be supplied by a wind from the secondary star, especially in the case of some recurrent novae with evolved secondaries like RS Oph (Bode et al. 2006; Sokoloski et al. 2006). In other cases, the ejecta may plow into ambient gas that results from an accretion disk wind (Drew & Proga 2000).

It is the final case, the soft X-ray emission from the white dwarf, that is of particular relevance to our discussion of nova populations in M31. The detection of GQ Mus (Nova Mus 1983) and later several other Galactic novae as super-soft [X-ray] sources (SSS) established that at least some novae go through an extended phase of residual hydrogen burning on the surface of the white dwarf after eruption (Ogelman et al. 1985; MacDonald & Vennes 1991). Early in the eruption the soft X-ray emission is absorbed by the ejecta. However once the ejecta become sufficiently optically thin, the hydrogen-burning shell on the white dwarf becomes visible, and the nova is seen as a SSS until the residual hydrogen is exhausted (Starrfield 1989; Krautter 2002). Surprisingly, follow-up studies of additional Galactic novae revealed only a few additional soft-X-ray emitting systems (Orio et al. 2001). It wasn't clear whether the soft X-ray phase was uncommon among novae, or whether the phase was simply missed due to its limited duration, to sparse X-ray coverage, or both. X-ray observations of novae in M31 would become key to advancing our understanding.

Observations of Novae in M31

For many of the same reasons that it has been ideal for the study of optical emission from novae, the Andromeda galaxy has been a major target for the study of X-ray emission from novae (Pietsch 2010). The high nova rate assures that there are dozens of novae discovered every year, the nova sample is essentially equidistant, making it possible to study relative nova properties without the need to determine distances individually, and, importantly for the X-ray observations, the foreground Galactic

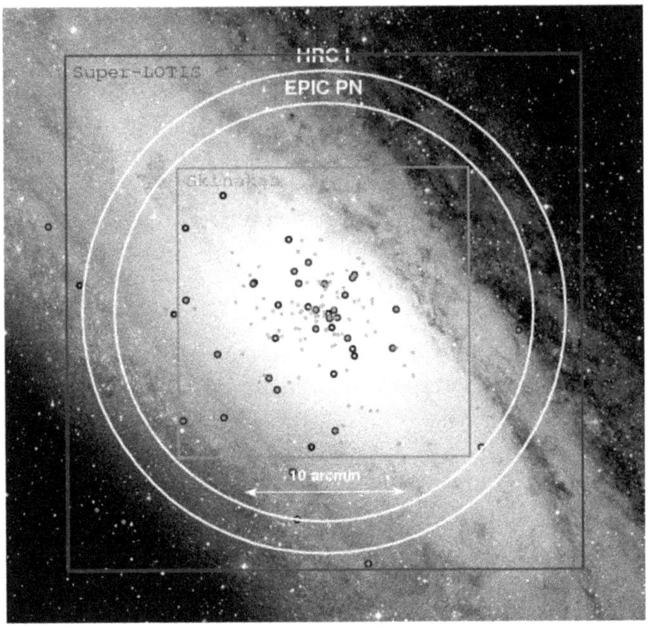

Figure 6.8. A Digital Sky Survey image of M31 showing the positions of novae in M31 from 2000 to 2010 (green dots) and the positions of X-ray detected novae (circles). Also shown are the fields of view of two optical (SuperLOTIS and Skinakas) and two X-ray instruments (*Chandra* HRC I and XMM-Newton EPIC PN) that have been key to our understanding of novae in M31. (From Pietsch 2010; copyright Wiley-VCH Verlag GmbH & Co. KGaA. Reproduced with permission.)

and internal M31 extinctions are low, making it feasible to study soft X-ray emission (see Figure 6.8).

In 2005, Wolfgang Pietsch and collaborators published the first of what would become a series of seminal papers on the study of the supersoft phase of novae in M31 (Pietsch et al. 2005). In their ground-breaking paper Pietsch et al. established that a SSS phase is common in post-eruption novae, and that novae represent the dominant population of SSS in M31. At about the same time, Sala & Hernanz (2005), and, shortly thereafter, Hachisu & Kato (2006, 2010) published studies of the X-ray emission from post-outburst novae and demonstrated that the X-ray properties could be used to constrain the fundamental physical properties of the progenitor binary, such as the white dwarf mass and chemical composition. With this potential in mind, Martin Henze et al. (2010, 2011, 2014), working with Pietsch and collaborators would go on to establish several important relationships between the soft X-ray behavior of post-novae (turn-on and turn-off times) and observed optical properties such as fade rate and shell expansion velocity, both of which are sensitive to system parameters such as the white dwarf mass and accretion rate (see Figure 6.9).

The SSS turn-on ($t_{\rm on}$) and turn-off ($t_{\rm off}$) times are positively correlated as expected, with the longest times being associated with slower novae with lower

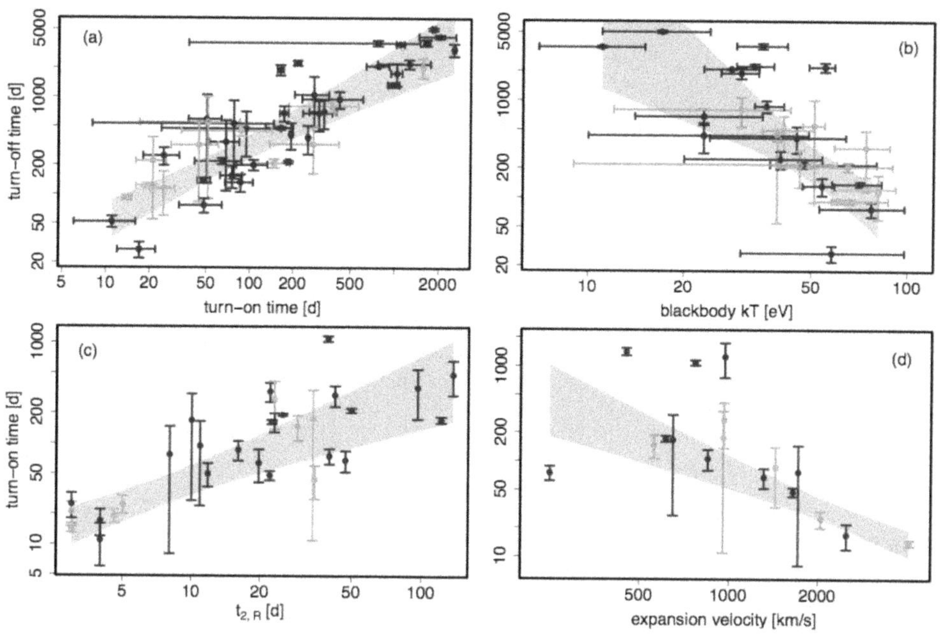

Figure 6.9. Correlations between soft X-ray turn-on and turn-off times and other nova parameters of interest. Blue and orange points represent novae from the disk and bulge samples, respectively. The green points represent novae discovered in globular clusters. (Reproduced with permission from Henze et al. 2014, © ESO.)

expansion velocities. Presumably such novae arise from lower mass white dwarfs that must accrete larger envelopes before the TNR is initiated. Similarly, t_{on} is inversely correlated with both the optical decline rate (positively correlated with t_2), and the shell expansion velocity. In both cases, a higher white dwarf mass will result in a lower ignition mass, which in turn will result in less material being ejected in the eruption, and thus shorter t_2 and t_{on} times.

An important aspect of the Henze et al. series was the attempt to use the supersoft X-ray properties to explore the question of whether two distinct populations of novae (bulge and disk) exist in M31. Although M31 is in many ways ideal for such a study, the high inclination of the galaxy to our line of sight makes an unambiguous separation of the populations difficult as foreground disk novae are projected onto the bulge. Nevertheless, Henze et al. (2011, 2014) assumed that bulge novae are mostly confined to an elliptical region with a semimajor axis of 700″, an ellipticity of 0.5, and a position angle of 50°, as shown in Figure 6.10. Novae falling outside this region were assumed to be novae associated with M31's disk. There was a slight suggestion that the disk novae had a higher mean blackbody temperature compared with the bulge sample, but the result was not statistically significant. However, a comparison of the cumulative distributions of novae with "high mass" white dwarfs (defined as novae with t_{on} < 100 d) with their "low mass" (t_{on} > 500 d) counterparts revealed that the former were more spatially extended compared with the latter

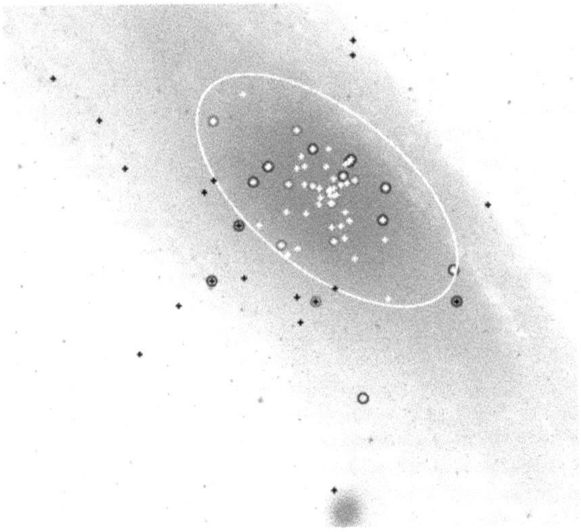

Figure 6.10. The positions of bulge (white) and disk (black) novae in M31. The ellipse delineates the assumed extent of M31's bulge. Circles indicated novae with high mass (blue) and low mass (red) white dwarfs as determined by their SSS turn-on times. (Reproduced with permission from Henze et al. 2014, © ESO.)

systems, and thus more likely associated with M31's disk population. This result is consistent with that found by Shafter et al. (2011) when comparing the optical fade rates of novae in M31—faster novae ($t_2 < 25$ d) were slightly more extended than their slower counterparts.

6.6.2 Infrared Observations of Novae

The study of Galactic novae has established that infrared observations can provide a wealth of information on the chemical abundance of nova ejecta, the process of dust formation, and the composition of the underlying white dwarf. (e.g., see Evans & Gehrz 2012; Gehrz et al. 1998 for reviews). One of the most celebrated novae of the twentieth century was Nova Herculis 1934 (DQ Her), which reached $m = 1.3$ at the peak of its eruption (Florja 1934). As described in Chapter 4, the nova would go on to play a prominent role in our realization that nova progenitors are all close binary systems. A peculiar feature of nova DQ Her that was perplexing at the time was a deep, nearly 5 mag dip in its optical light curve commencing ~100 days after eruption and persisting until approximately day 200 (see Figure 6.11). An explanation of this feature would have to wait another 35 years before infrared observations of a similar nova would solve the mystery. In particular, infrared photometry of FH Ser (Nova Serpentis 1970) revealed a strong infrared excess ~2 months after the eruption produced by dust formation in the nova's ejecta (Geisel et al. 1970; Hyland & Neugebauer 1970; Gallagher 1977). Studies of additional novae revealed that not all novae form significant amounts of dust. The copious carbon dust produced in novae such as DQ Her and FH Ser can only result from

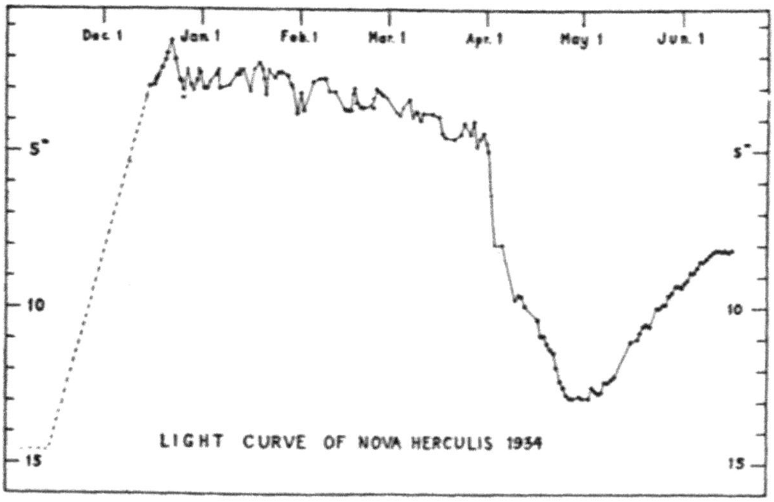

Figure 6.11. The optical light curve of Nova Herculis 1934 (DQ Her) showing the prominent dust dip starting ~100 days after eruption. (Reproduced with permission from Wyse 1935, © ESO.)

thermonuclear runaways on CO white dwarfs (Gehrz et al. 1998; José et al. 2016). On the other hand, novae arising from ONe white dwarfs do not form carbon dust, but may produce modest amounts of silicates.

Another of the early discoveries made possible by infrared observations was the detection in 1985 of strong [Ne II] 12.8 μm emission in the Galactic nova QU Vul (Nova Vulpeculae 1984#2) (Gehrz et al. 1985). The line, which peaked at a flux a factor of ~60 above the free–free continuum, was the strongest [Ne II] line relative to the continuum ever seen in any astrophysical source (see Figure 6.12). Several additional novae displaying extremely strong forbidden lines of neon would be recognized in subsequent years (e.g., V1370 Aql and V693 CrA, Andrea et al. 1994), and would become collectively known as "Neon Novae." The abundances of neon in these novae are sufficiently high (one to two orders of magnitude above solar abundance) that it must be the result of dredge-up from an ONe white dwarf (Casanova et al. 2016; Glasner et al. 2012; Gehrz et al. 2008; Iliadis et al. 2002). This has important implications for the role of novae as progenitors of SNe Ia as accreting ONe white dwarfs are expected to undergo accretion-induced collapse upon reaching the Chandrasekhar limit (Miyaji et al. 1980; Schwab et al. 2015). Rather, it is the accreting CO white dwarfs that are capable of detonating as type Ia supernovae. Clearly, a knowledge of the relative populations of CO and ONe novae will be crucial to evaluating the potential of novae as a viable channel for the production of SNe Ia.

An IR Survey of Novae in M31

The first infrared survey of novae outside the Galaxy was undertaken by Shafter et al. (2011) in 2011 using the *Spitzer Space Telescope*. A total of 10 novae were

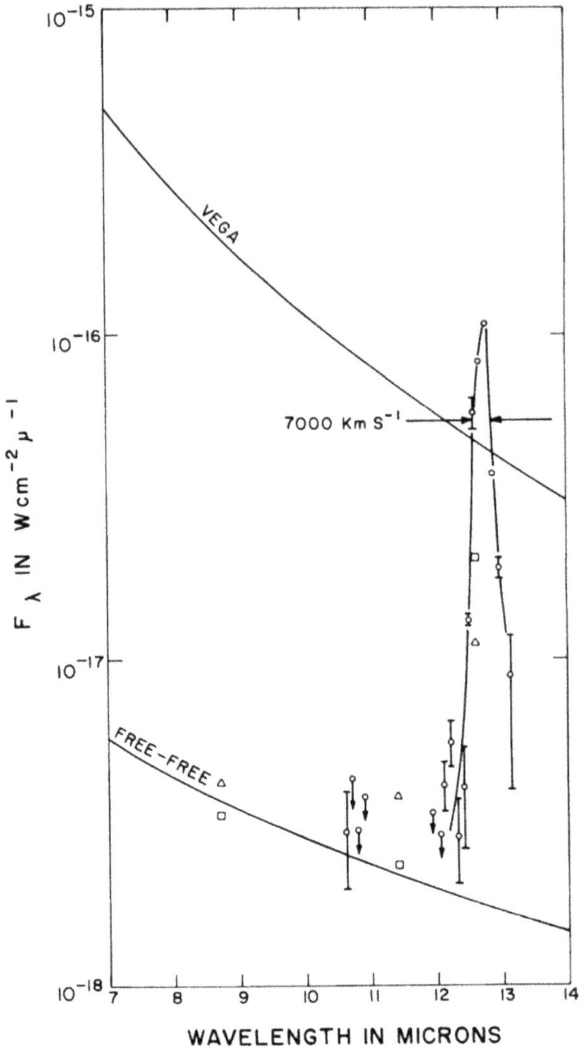

Figure 6.12. The infrared spectrum of QU Vul showing the strong [Ne II] 12.8 μm emission line. (Reproduced from Gehrz et al. 1985. © 1985. The American Astronomical Society. All rights reserved.)

observed in M31—eight with IRAC (Infra-Red Array Camera) and eight with IRS (Infra-Red Spectrometer), with a total of six novae being observed with both instruments. Figure 6.13(a) shows the near-infrared energy distributions for the six novae detected with IRAC. Although modest in scope, the survey revealed one dust-forming nova (M31N 2006-10a) and possibly a second (M31N 2007-07f). In the case of 2006-10a a total dust mass of $\sim 2 \times 10^{-6}$ $M_\odot$ was deduced, which is comparable to Galactic novae that form extensive dust shells. In addition to the IRAC results, the IRS observations revealed strong [Ne II] emission in a third nova (M31N 2007-11e) raising the possibility that this object is a Neon nova (see Figure 6.13(b)).

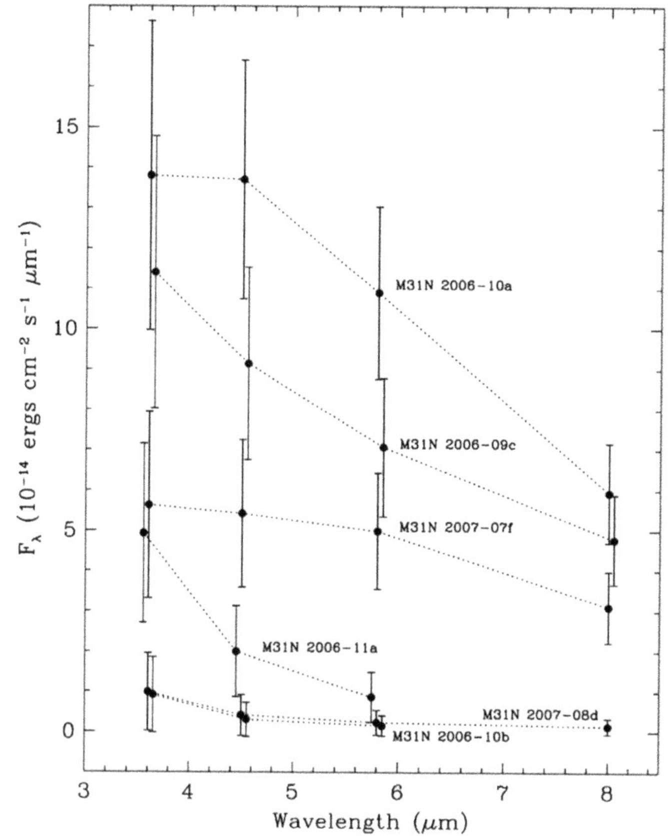

(a)

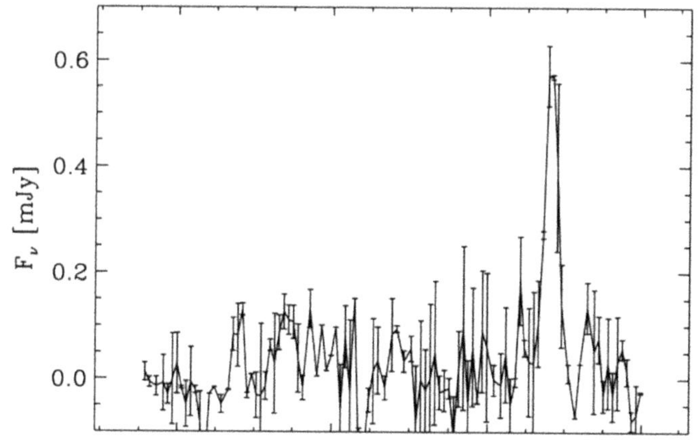

(b)

Figure 6.13. (a) IRAC energy distributions for six novae in M31. Note the infrared excess in M31N 2006-10a and possibly 2007-07f. (b) IRS spectrum of M31N 2007-11e showing the strong [Ne II] 12.8 μm emission. (Reproduced from Shafter et al. 2011. © 2011. The American Astronomical Society. All rights reserved.)

A major limitation of the M31 survey was the long delay between the time of an eruption and the scheduling of the *Spitzer* observations. As a result it is possible that some of the targets may have formed dust before the observations were initiated. Despite the limitations, the study demonstrated the potential that future IR surveys of M31, such as those possible with the *James Webb Space Telescope,* will have on our understanding of the role novae play in the chemical evolution of galaxies, and as potential SNe Ia progenitors.

References

Andrea, J., Drechsel, H., & Starrfield, S. 1994, A&A, 291, 869
Arp, H. C. 1956, AJ, 61, 15
Bode, M. F., O'Brien, T. J., Osborne, J. P., Page, K. L., & Senziani, F. 2006, ApJ, 652, 629
Casanova, J., José, J., Garca-Berro, E., & Shore, S. N. 2016, A&A, 595, 28
Ciardullo, R., Ford, H., & Jacoby, G. 1981, BAAS, 13, 842
Ciardullo, R., Ford, H., & Jacoby, G. 1983, ApJ, 272, 92
Ciardullo, R., Ford, H. C., Neill, J. D., et al. 1987, ApJ, 318, 520
Cowley, A. P., & Starrfield, S. G. 1987, PASP, 99, 854
Crayton, H., Rector, T. A., Walentosky, M. J., et al. 2018, AAS, 232, 320.08
Darnley, M. J., Bode, M. F., Kerins, E., et al. 2004, MNRAS, 353, 571
Darnley, M. J., Bode, M. F., Kerins, E., et al. 2006, MNRAS, 369, 257
Dopita, M. A., Rosino, L., & Dodorico, S. 1979, A&A, 76, 240
Drew, J. E., & Proga, D. 2000, NewAR, 44, 21
Evans, A., & Gehrz, R. D. 2012, BASI, 40, 213
Florja, N. 1934, TsTas, 36, 1
Ford, H., Ciardullo, R., & Jacoby, G. 1981, IAU Circ., No. 3643, #1,
Ford, H., Jacoby, G., Ciardullo, R., Ulrich, R., & Shafter, A. 1982, IAU Circ., No. 3725, #3,
Gallagher, J. S. 1977, AJ, 82, 209
Gehrz, R. D., Grasdalen, G. L., & Hackwell, J. A. 1985, ApJ, 298, 47
Gehrz, R. D., Truran, J. W., Williams, R. E., & Starrfield, S. 1998, PASP, 110, 3
Gehrz, R. D., Woodward, C. E., Helton, L. A., et al. 2008, ApJ, 672, 1167
Geisel, S. L., Kleinmann, D. E., & Low, F. J. 1970, ApJ, 161, 101
Glasner, S. A., Livne, E., & Truran, J. W. 2012, MNRAS, 427, 2411
Hachisu, I., & Kato, M. 2006, ApJS, 167, 59
Hachisu, I., & Kato, M. 2010, ApJ, 709, 680
Henze, M., Pietsch, W., Haberl, F., et al. 2010, A&A, 523, 89
Henze, M., Pietsch, W., Haberl, F., et al. 2011, A&A, 533, 52
Henze, M., Pietsch, W., Haberl, F., et al. 2014, A&A, 563, 2
Hubble, E. 1929, ApJ, 69, 103
Humason, M. 1932, PASP, 44, 381
Hyland, A. R., & Neugebauer, G. 1970, ApJ, 160, 177
Iliadis, C., Champagne, A., José, J., Starrfield, S., & Tupper, P. 2002, ApJS, 142, 105
José, J., Halabi, G. M., & El Eid, M. F. 2016, A&A, 593, 54
Kasliwal, M. 2011, PhD thesis, California Institute of Technology
Kasliwal, M. 2012, PASA, 29, 482
Krautter, J. 2002, in AIP Conf. Ser. 345, Classical Nova Explosions, ed. M. Hernanz (New York: AIP)

Law, N. M., Kulkarni, S. R., & Dekany, R. G. 2009, PASP, 121, 1395
MacDonald, J., & Vennes, S. 1991, ApJ, 373, 51
Miyaji, S., Nomoto, K., Yokoi, K., & Sugimoto, D. 1980, PASJ, 32, 303
Mukai, K. 2017, PASP, 129, 976
Ögelman, H., Beuermann, K., & Krautter, J. 1985, ESA Recent Results on Cataclysmic Variables, (Noordwijk: ESA Scientific and Technical Publications Branch)
Orio, M., Covington, J., & Ögelman, H. 2001, A&A, 373, 542
Patterson, J., & Raymond, J. C. 1985a, ApJ, 292, 535
Patterson, J., & Raymond, J. C. 1985b, ApJ, 292, 550
Pietsch, W., Fliri, J., Freyberg, M. J., et al. 2005, A&A, 442, 879
Pietsch, W. 2010, AN, 331, 187
Rau, A., Kulkarni, S. R., & Law, N. M. 2009, PASP, 121, 1334
Rosino, L. 1964, AnAp, 27, 498
Sala, G., & Hernanz, M. 2005, A&A, 439, 1061
Schwab, J., Quataert, E., & Bildsten, L. 2015, MNRAS, 453, 1910
Shafter, A. W., & Irby, B. K. 2001, ApJ, 563, 749
Shafter, A. W., Darnley, M. J., Hornoch, K., et al. 2011, ApJ, 734, 12
Shafter, A. W., Henze, M., Rector, T. A., et al. 2015, ApJS, 216, 34
Sokoloski, J. L., Luna, G. J. M., Mukai, K., & Kenyon, S. J. 2006, Natur, 442, 276
Soraisam, M. D., Gilfanov, M., Kupfer, T., et al. 2017, A&A, 599, 48
Starrfield, S. 1989, Classical Novae, ed. M. Bode, & A. Evans (New York: Wiley), 39
Tomaney, A. B., & Shafter, A. W. 1992, ApJS, 81, 683
Tomaney, A. B., & Shafter, A. W. 1993, ApJ, 411, 640
Williams, R. E. 1992, AJ, 104, 725
Wyse, A. B. 1935, ASPL, 2, 129 83

Chapter 7

Extragalactic Nova Rates and Populations

Most of what we have learned about nova properties across differing stellar populations has come from extragalactic surveys. As we saw in the preceding chapter, Galactic novae, although closer and brighter, are generally more difficult to study given that they are preferentially located close to the Galactic plane and toward the bulge where the variable obscuration caused by interstellar dust makes estimates of distances and luminosities challenging.

We will begin by discussing what we have learned from observations of novae in the local group, but first it is appropriate to describe how nova rates are determined in extragalactic systems.

7.1 Measuring Extragalactic Nova Rates

7.1.1 The Control Time Method

Many of the early extragalactic nova rate estimates have made use of the "control time" methodology first used by Zwicky (1942) in his work on estimating supernova rates from synoptic surveys. As applied to novae, the rate in the surveyed region of a given galaxy is determined from the ratio of the observed number of novae, $N_{\rm obs}$ and the survey's "control time," $T_{\rm c}$. The control time is the *total* amount of time during the course of the survey where an *average* nova would remain brighter than the survey's limiting magnitude, $m_{\rm lim}$. For a survey consisting of n epochs of observation at times, t_i, where the average nova remains brighter than $m_{\rm lim}$ for a time $\tau_{\rm c}$, the control time is given by:

$$T_{\rm c} = \tau_{\rm c} + \sum_{i=2}^{n} \min(\tau_{\rm c},\, t_i - t_{i-1}). \qquad (7.1)$$

The nova rate in the surveyed region of a galaxy is then simply given by:

doi:10.1088/2514-3433/ab2c63ch7

$$R = \frac{N_{obs}}{T_c}. \tag{7.2}$$

A limitation of the control time method is that the parameter, τ_c, the mean nova lifetime brighter than the survey limit, must be specified. This, in turn, requires an *a priori* knowledge of the light curve properties (peak luminosity and fade rate) of the underlying nova population. A common practice is to estimate τ_c from an equidistant sample of nova light curves from a nearby galaxy with a high nova rate, such as M31. The dangers in this approach are twofold: (1) Unless the nova survey of the target galaxy has a cadence and depth comparable to the survey employed to determine τ_c, selection effects may possibly bias the observed nova sample against the detection of fainter and faster novae, and (2) there is no guarantee that the population of novae in a given galaxy under study will possess the same light curve characteristics as those in the sample (e.g., from M31) used to determine τ_c.

7.1.2 Numerical Simulations

Extragalactic nova rates can also be estimated through numerical simulations designed to determine the annual nova rate required to produce the number of novae discovered in a given survey. Specifically, for a given assumed nova rate, R, the number of novae we would expect to discover in a given survey, N_{obs}, is a function of the survey's completeness as a function of magnitude, $C(m)$, the frequency of observation, and the properties of the nova light curves. To begin, a set of model light curves is computed based on randomly selecting peak magnitudes and decay rates from a suitable sample of light curves thought to be representative of the nova population in the target galaxy. Then, based on the survey dates and the adopted distance to the target galaxy, a model nova luminosity function, $n(m, R)$, is computed. The resulting luminosity function is then convolved with a survey completeness function, $C(m)$, to yield the expected number of observed novae, N_{obs}:

$$N_{obs}(R) = \sum_m C(m)\ n(m, R). \tag{7.3}$$

The computation is repeated multiple times and the number of matches between $N_{obs}(R)$ and the actual number of novae observed in the survey, n_{obs}, is recorded. When normalized by the number of trials, the number of matches as a function of R produces a probability function for the galaxy's nova rate.

Like the control time method, numerical simulations of nova rates in extragalactic systems require that the light curve properties of the nova population be specified *a priori*. In particular, for the case of numerical simulations it is the model luminosity function—the number of novae visible at a given time as a function of magnitude—that must be specified. As for the control time method, any uncertainty in the assumed nova light curve properties will be propagated to the nova rate determination.

7.1.3 Luminosity-specific Nova Rates

In order to compare nova rates between different galaxies and across differing stellar populations, the rates must be suitably normalized. It would be ideal to normalize the nova rates by the mass (in stars) of the galaxy under study, however galaxy masses are notoriously difficult to measure. As a proxy for the galaxy mass, the standard practice since the work of Ciardullo and collaborators in the early 1990s (Ciardullo et al. 1990a) has been to normalize extragalactic nova rates by the infrared *K*-band luminosity of the galaxy.[1] The near-infrared light, as opposed to the blue or visible light, is less sensitive to the disproportionately high luminosity contributed by a small number of massive stars, and thus provides a better tracer of the overall mass in stars.

As a practical matter, the infrared luminosity of a galaxy of known distance can be measured directly using wide-field infrared imaging, or it can be estimated through optical imaging coupled with an estimate of the galaxy's infrared color, e.g., $V–K$. Early estimates of a galaxy's *K*-band luminosity-specific nova rate ($LSNR_K$) were made using infrared colors, since direct measurements of a galaxy's integrated infrared light were usually not available. The situation changed with the introduction of the Two-Micron All-Sky Survey (2MASS) in 2003 (Jarrett et al. 2003). Unfortunately, performing accurate background subtraction for galaxies with the largest angular sizes—typically closest galaxies with measured nova rates—proved difficult, and thus the resulting 2MASS magnitudes were often unreliable. For example, in the case of M31 and M33, the 2MASS integrated *K*-band magnitudes significantly underestimate the actual fluxes (Jarrett 2012, private communication).

7.2 Nova Rates in the Local Group

According to the online nova database maintained by Pietsch (2010), until the end of 2018 a total of 48 novae had been discovered in M33, 54, and 22 in the LMC and the Small Magellanic Cloud (SMC), respectively, four each in M32 and NGC 205, and roughly 1100 in M31. To these we can add 2 and 3 novae, respectively, in the dwarf irregulars NGC 6822 and IC 1613, and almost 500 nova candidates that have been reported in the Milky Way over the past ~400 years. The data on local group galaxies is summarized in Table 7.1. Nova production in the local group is dominated by the spiral galaxies M31 and the Galaxy, with M33 and the Magellanic Clouds playing relatively minor roles. Although novae have been discovered in the dwarf elliptical galaxies NGC 205 (M110) and NGC 221 (M32), and in the dwarf irregular system NGC 3109, the number of novae detected is too low for meaningful estimates of their nova rates to be made.

[1] The early work of Della Valle et al. (1994) used infrared *H*-band luminosities to normalize their nova rate estimates.

Table 7.1. Novae in Local Group Galaxies

Galaxy	Number of novae	Estimated rate (per year)	Reference
M31	~1100	65^{+16}_{-15}	Pietsch (2010); Darnley et al. (2006)
M33	48	$2.5^{+1.0}_{-0.7}$	Williams & Shafter (2004)
M32	4	…	Pietsch (2010)
LMC	54	2.4 ± 0.8	Pietsch (2010); Mróz et al. (2016)
SMC	22	0.9 ± 0.4	Pietsch (2010); Mróz et al. (2016)
NGC 205	4	…	Pietsch (2010)
NGC 3109	1	…	Brimacombe et al. (2018)
NGC 6822	2	…	Healy et al. (2019); King & Li (1999); Wei et al. (1999)
IC 1613	3	…	Sandage (1971); Mantegazza et al. (2001); Williams et al. (2015)
Milky Way	~400	50^{+31}_{-23}	Shafter (2017)

7.2.1 M31

Although the discovery of novae in the local group dates back a century in most cases, the majority of novae have been discovered in recent years. In the case of M31 specifically, approximately half of the more than 1000 novae thus far identified have been discovered during the past 20 years, with many being contributed by amateur astronomers. Table 7.2 summarizes the breakdown of major nova surveys for in M31 over the past century. Many of theses surveys have been discussed in some detail in previous chapters; here, we will simply summarize what we have learned from these surveys regarding the rates and properties of novae in M31.

The Observed M31 Nova Rate

The first estimate of the nova rate in M31 was made by Hubble in his 1929 paper (Hubble 1929). Based on the total of 85 novae known at the time, Hubble estimated an annual nova rate of ~30 per year. Remarkably there was little change in this estimate over the next half century, with Arp finding a rate of 26 ± 4 yr^{-1} (Arp 1956), Rosino in 1964 estimating 29 per year (Rosino 1964), and Capaccioli et al. who analyzed the complete sample of novae from Rosino's study, confirming the preliminary rate of 29 ± 4 novae per year (Capaccioli et al. 1989). It is important to note that all of these estimates were based upon novae detected in photographic surveys. Sophisticated image differencing techniques that are now possible in working with modern digital images are better able to detect fainter novae, especially in the bright inner bulge region of M31 where the nova density is high. This increased sensitivity has resulted in generally higher nova rate estimates. Using the control time method, Shafter & Irby (2001) estimated a rate of 37^{+12}_{-8} per year, while Darnley et al. (2006) found an even higher rate of 65^{+16}_{-15} per year based on

Table 7.2. Major Surveys for Novae in M31

Survey	Dates	Number	Obs/Telescope	Reference
Hubble	1909–1927	85	Mt Wilson 60 and 100 in Plates	Hubble (1929)
Arp	1953–1955	30	Mt Wilson 60 in plates	Arp (1956)
Rosino	1955–1963	46	Asiago 122 cm reflector	Rosino (1964)
Rosino	1963–1970	44	Asiago 122 cm, 92-67 Schmidt	Rosino (1973)
Sharov & Alksnis	1969–1989	21	Crimea, Latvia, B plates	Sharov & Alksnis (1991)
Ciardullo et al.	1982–1989	40	KPNO 0.9 m, McD 0.8 m, Wise 1 m	Ciardullo et al. (1987)
Rosino et al.	1971–1986	52	Asiago	Rosino et al. (1989)
Shafter & Irby	1990–1997	82	Mt Laguna 40 in	Shafter & Irby (2001)
RBSE	1996–2017	71	KPNO 0.9 m	Crayton et al. (2018)
Darnley et al.	1999–2002	20	POINT-AGAPE	Darnley et al. (2004, 2006)
Hornoch	2002–2017	185	Ondřejov Observatory	Hornoch (2018, private communication)

numerical simulations. Population synthesis models have predicted a substantial population of faint and rapidly-fading population of novae many of which may have been missed in earlier surveys. If so, the nova rate in M31 required to generate the observed nova population may approach 100 per year (Chen et al. 2016).

The corresponding LSNR_K for M31 is even more uncertain, as it also depends on the adopted K-band luminosity of the Galaxy. As described earlier, the 2MASS magnitude for M31, $K = 0.984$, is unreliable and is significantly fainter than what one estimates based on M31's V magnitude and $V–K$ color. If we adopt a total, extinction-corrected visual magnitude, $V_{T,0} = 2.68$ for M31 (de Vaucouleurs et al. 1991), coupled with a $V–K$ color characteristic of M31's bulge $V–K = 3.38$ (Sandage et al. 1969) we find $K = -0.7$ (almost a factor of 5 brighter than the published 2MASS K-band flux!). Given $K_\odot = 3.27$ (Willmer 2018), a foreground extinction toward M31 of $A_K = 0.0$ (Schlegel et al. 1998), and a distance to M31 of $\mu_0 = 24.38$ (Freedman et al. 2001), we find $L_{K,M31} = 2.23 \times 10^{11}\ \mathrm{L}_{\odot,K}$. For a nova rate of $65 \pm 15\ \mathrm{yr}^{-1}$ this corresponds to an M31 LSNR_K of 2.9 novae per year per 10^{10} solar luminosities in K.

It is likely that this is a lower limit to the true LSNR_K for M31, both because the nova rate may be underestimated, and because spiral galaxies like M31 become bluer with increasing distance from the Galaxy's center. This suggests that the appropriate $V–K$ color for M31 as a whole will be bluer than that of M31's bulge. If we adopt an integrated color characteristic of Sb and Sbc galaxies, $V–K \simeq 3.17$, from Aaronson's early work (Aaronson 1978) on the infrared colors of bright nearby galaxies, we find a modified LSNR_K of 3.5 novae per year per 10^{10} solar luminosities in *K*. Based on the available data, we consider this value to represent the best estimate currently available for the *K*-band luminosity-specific nova rate of M31.

The Population of M31 Novae

It has been recognized since Hubble's original survey that novae in M31 appear to generally follow the light distribution of the galaxy, and thus are strongly concentrated towards the nuclear bulge region. When considering the distribution of novae along the major axis of M31, Arp (1956) determined that "[t]he frequency of novae drops off with increasing radius more sharply than the general light of the nebula." As described in the previous chapter, the spatial distribution of novae was further quantified by Ciardullo et al. (1987) who concluded that the nova density per unit light in the bulge was at least 10 times that in the disk, and that the data were consistent with *all* of the novae coming from M31's bulge population. The strong association of M31's novae with the galaxy's bulge population was confirmed by the subsequent studies of Shafter & Irby (2001) and Darnley et al. (2006).

Not all studies supported the notion that M31's nova population was bulge dominated. In 1997, Kazuhito Hatano and collaborators introduced a new and interesting approach to the study of nova populations in M31 (Hatano et al. 1997). Specifically, they realized that given the high inclination of M31 to our line of sight ($i \sim 77°$), the spatial distribution of novae across the apparent bulge of M31 (which includes disk novae projected onto the bulge) will be affected by extinction in M31's disk. By comparing the observed distribution of novae with those produced by various model distributions, Hatano et al. found that bulge-dominated nova models produced a poor match to the observations. Alternatively, the best fit was achieved when model novae were assumed to arise equally from the disk and bulge populations.

In yet another approach to the study of nova populations, starting in 2006 Shafter and collaborators initiated long-term spectroscopic and photometric survey of novae in M31 using the HET at McDonald Observatory (Shafter et al. 2011). Over the course of four years a total of 53 spectra of 48 individual novae in M31 were obtained, which effectively doubled the number of M31 nova spectra available, bringing the total to 91 spectra by the start of 2010. The primary goal was to obtain spectra suitable for classification of the nova as either belonging to the Fe II, He/N or Hybrid classes (Williams 1992). The spectroscopic survey was augmented with follow-up photometry of 47 novae sufficient to determine maximum magnitudes and rates of decline.

As covered in the last chapter, earlier work by Valle & Livio (1998) had shown that for Galactic novae the He/N novae were clustered closer to the Galactic plane compared with the Fe II systems. Their interpretation was that the He/N novae were primarily associated with the Galactic disk population, while the Fe II novae were associated with the thick-disk and bulge. The Galactic study had also revealed that novae classified as He/N were on average more luminous and faster declining than their Fe II counterparts. These results provide strong support for the two-nova-population hypothesis, and it would be nice to confirm the findings in M31. With this goal in mind, a major focus of the HET spectroscopic survey was to

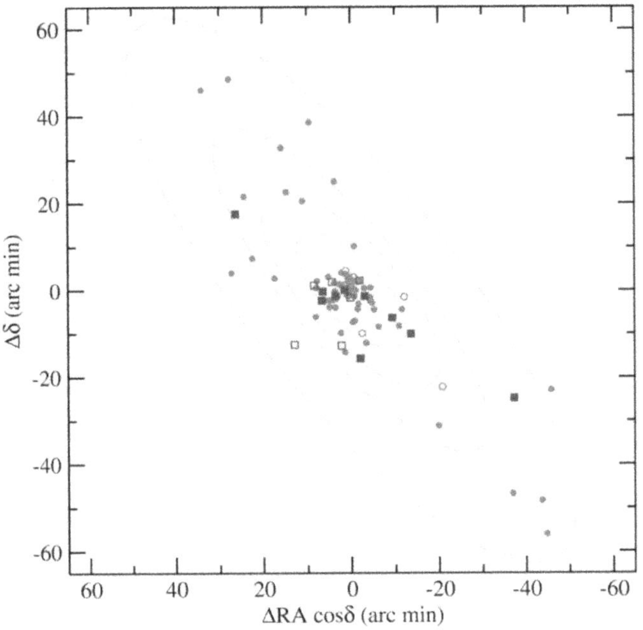

Figure 7.1. The spatial distributions of the Fe II and He/N novae in M31 from the HET survey of Shafter et al. (2011). Filled blue squares: He/N novae; open blue squares: uncertain He/N novae; filled red circles: Fe II novae; open red circles: uncertain Fe II novae. (Reproduced from Shafter et al. 2011. © 2011. The American Astronomical Society. All rights reserved.)

determine whether the relative fraction of novae with different spectroscopic classes (Fe II and He/N) varied spatially within M31.

Figure 7.1 shows the spatial distribution of the Fe II and He/N novae in M31 from the HET survey. Most of the novae are concentrated in the bulge region where the nova rate is highest, making it difficult to study the overall spatial distribution. Nevertheless, there doesn't appear to be any obvious difference in the overall spatial distributions between the two spectroscopic classes. This impression is confirmed by the cumulative distributions shown in Figure 7.2. A Kolomogorov–Smirnov (K–S) test finds no statistically significant difference between the distributions for Fe II and He/N novae (K–S statistic ~0.8).

The nova speed classes were also used to explore possible differences between the putative bulge and disk populations. Novae were divided into two groups—a "fast" group with $t_2 < 25$ d and a "slow" group with t_2 times of 25 d or longer. The cumulative distributions are shown in Figure 7.3. A K–S statistic of 0.23 suggests that the two distributions differ with the fast novae being slightly more spatially extended, but the result is not statistically significant.

Shortly after the publication of the Shafter et al. spectroscopic survey, Steven Williams and collaborators took a different approach to the study of nova populations in M31, initiating a program to explore the progenitors of M31 novae using archival *Hubble Space Telescope* (*HST*) observations (Williams et al. 2014a, 2014b). From a total of 38 novae with *HST* coverage, progenitors for 11 systems

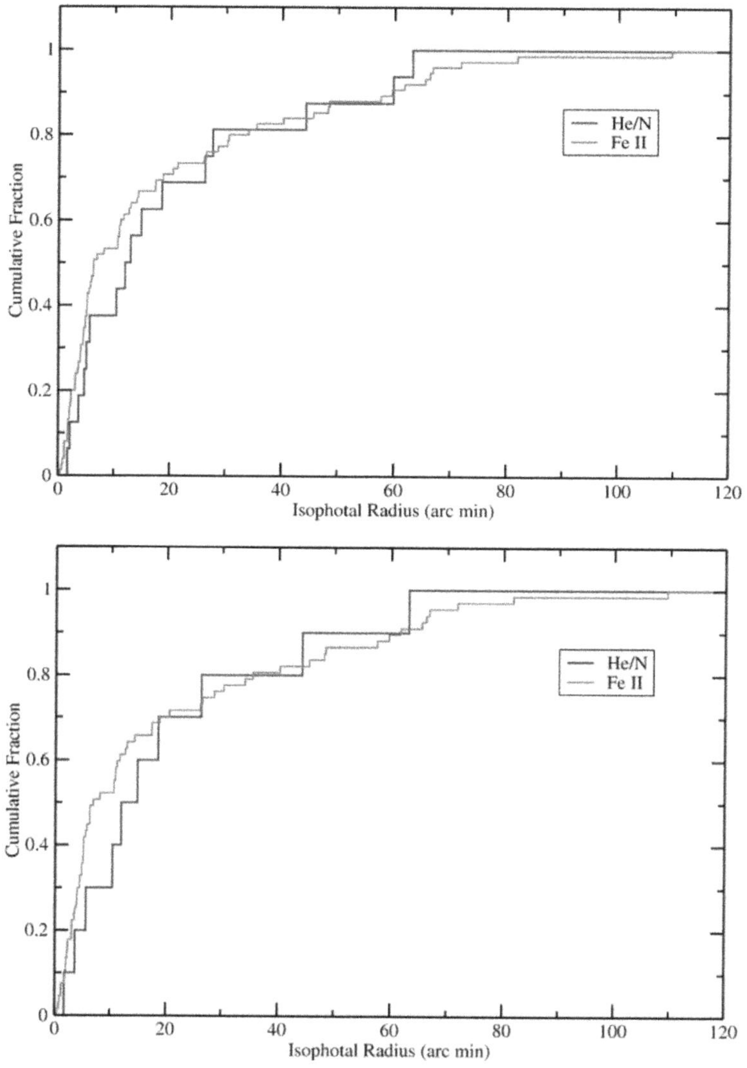

Figure 7.2. Cumulative distributions of the Fe II (red) and He/N (blue) novae in M31. (a) All novae, including those where the Fe II and He/N classification is uncertain; (b) Only novae with certain spectroscopic classifications. (Reproduced from Shafter et al. 2011. © 2011. The American Astronomical Society. All rights reserved.)

were recovered leading to the conclusion that up to 24% of the M31 novae studied contain evolved secondary stars. A subsequent and more detailed analysis revealed that 30^{13}_{-10} percent of the entire M31 nova population harbored giant secondaries. Furthermore, a population analysis showed that the red-giant novae were more likely to be associated with the disk of M31 (Williams et al. 2016). This is a remarkable result given that giant secondaries have been associated with only an estimated 3% of Galactic novae, most of which are located in the Galactic disk.

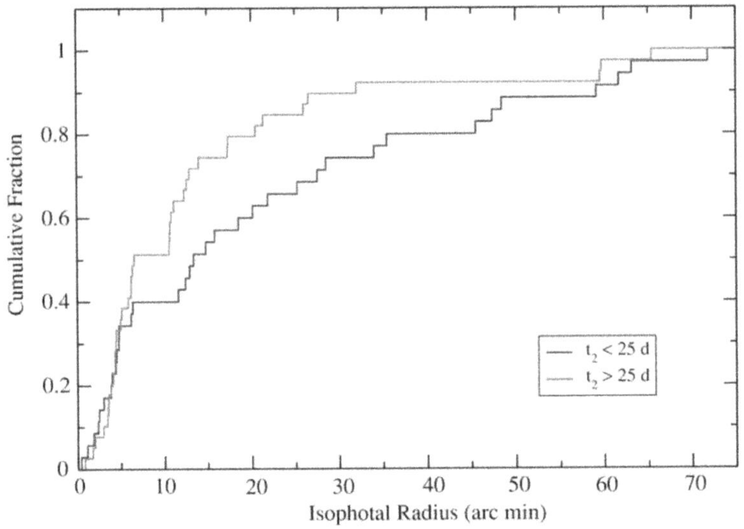

Figure 7.3. Cumulative distributions of "fast" ($t_2 < 25$ d) and "slow" ($t_2 > 25$ d) novae, blue and red lines, respectively (from the Shafter et al. M31 survey). A K–S test suggests that the distributions are possibly different with the fast novae being more spatially extended, but the difference is not statistically significant. (Reproduced from Shafter et al. 2011. © 2011. The American Astronomical Society. All rights reserved.)

In conclusion, the findings from the M31 HET survey failed to corroborate the results of Della Valle and Livio's study of Galactic novae. In addition, a very recent study by Özdönmez et al. (2018) of the spatial distribution of spectroscopic classes for a large sample of Galactic novae also failed to find any compelling evidence for two separate nova populations. On the other hand, the possibility that red-giant novae appear to be primarily associated with the disk population would tend to support the two-population hypothesis. Further observations, both Galactic and extragalactic, will be required before we can definitively determine whether there are two distinct populations of novae. The best evidence at present is that if two populations do indeed exist, the differences do not appear to be striking.

7.2.2 The Magellanic Clouds

It has been more than a century since the first nova was discovered in the Magellanic Clouds. Indeed, the first nova in the Clouds was observed in the late 19th Century and predates the first nova discovered in M31 by a dozen years. The nova, now designated as SMCN 1897-10a was originally discovered as a variable source by Henrietta Leavitt, and later confirmed as a nova by McKibbem (1951). The first nova recorded in the Large Magellanic Cloud (LMC), LMCN 1926-09a, was discovered on 1926 September 28 and subsequently announced by Luyten (1927).

The LMC and SMC Nova Rates

Early efforts to estimate the nova rates in the Magellanic Clouds were frustrated by their relatively low masses and large angular sizes (~100 deg^2 and 16 deg^2 for the

LMC and SMC, respectively). It also didn't help that the Clouds were located in the southern hemisphere where the observational resources were relatively limited until recently. An early attempt to quantify nova rates in the Clouds was made by Graham (1979) who estimated a rate of two to three per year in the LMC, with one nova every three to four years in the SMC. Later (Capaccioli et al. 1990) reanalyzed the available data and arrived at a slightly lower LMC nova rate of $\sim 2 \pm 1$ per year.

The first comprehensive study of novae in the Magellanic Clouds was made possible by the Optical Gravitational Lensing Experiment (OGLE) project. OGLE is a long-term project based at the University of Warsaw, and led by Andrej Udalski (Udalski & Kubiak 1997; Udalski 2008, 2015). Although designed primarily to detect microlensing events towards the rich stellar backgrounds provided by the Galactic bulge and the Magellanic Clouds, the project has also detected thousands of variable stars, both pulsating and eclipsing.

Incorporating data obtained over a time span of 16 years between 1997 and 2013, Przemek Mróz and collaborators detected a total of 20 novae in the LMC, 13 of which were new discoveries (Mróz et al. 2016). After correcting for temporal and spatial coverage, and detection efficiency, the authors arrived at an annual nova rate of 2.4 ± 0.8 yr^{-1}, in close agreement with the initial estimate made by Graham nearly 40 years earlier. In the case of the Small Cloud, a similar analysis of seven novae discovered in the SMC, yielded a nova rate of 0.9 ± 0.4 annually. After normalizing by the infrared luminosities of the Clouds, Mróz et al. determined K-band luminosity-specific nova rates, $\nu_K = 6.5 \pm 2.2$ and 7.5 ± 3.1 novae per year per 10^{10} solar luminosities in the K-band, for the LMC and SMC, respectively.

Nova Populations in the LMC

The properties of novae in the LMC was recently investigated by Shafter (2013). Of the 43 novae known at the time of the study, rates of decline were available for 28 systems and spectroscopic classes were available for 18. The rates of decline for the LMC nova sample were found to be considerably faster that those of M31 and the Galaxy (see Figure 7.4). In addition, the mix of spectroscopic classes of LMC novae appear to be different from those in M31 and the Galaxy, with approximately 50% belonging to the He/N and Fe IIb spectroscopic classes. Both the faster fade rates and the higher percentage of He/N and Fe IIb systems suggest that the LMC novae contain on average more massive white dwarfs than do their M31 and Galactic counterparts. This conclusion is also supported by the relatively high fraction of LMC novae that are recurrent (see Chapter 8).

7.2.3 M33

M33, classified as Hubble type SA(s)cd, is a nearby, low-mass, late-type spiral galaxy with a correspondingly modest nova rate. The first recorded nova in M33 (M33N 1919-12a) was discovered on a plate obtained by Pease with the Mount Wilson 100 inch reflector on 1919 December 14 (Hubble 1926). Three years later, in late 1922 August while studying three variable stars in M33, John Duncan noticed a previously unseen star of magnitude 17.5 on a plate taken with the Mount Wilson

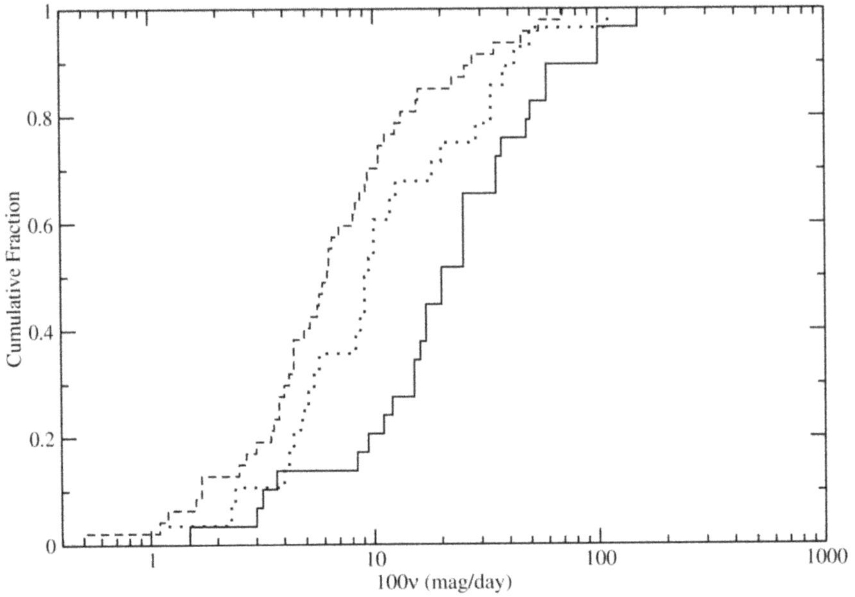

Figure 7.4. Cumulative rates of decline for LMC novae (solid line) compared with novae in M31 (dash) and the Galaxy (dots). The LMC novae are considerably "faster" than their M31 and Galactic counterparts. (Reproduced from Shafter 2013. © 2013. The American Astronomical Society. All rights reserved.)

60 inch reflector. No follow-up observations were obtained, but the object was likely the second nova discovered in M33. Over almost the next half century (through 2018) a total of 48 nova candidates have been observed in M33.

The M33 Nova Rate

The first estimates of the nova rate in M33 date back to the mid 1950s. During Arp's survey of M31, several additional Local Group galaxies were also imaged periodically. No novae were detected in M33, but based on the temporal coverage Arp (1956) was able to estimate an upper limit of $\sim$1 novae per year in that galaxy.

A total of five novae in M33 were discovered on 32 blue-sensitive blue plates taken by Rosino between 1960 and 1971 using the 122 cm telescope at Asiago (Rosino & Bianchini 1973). Given the sporadic temporal coverage, these observations did not permit a meaningful estimate of the nova rate; however, the available data suggested that the nova rate "seems not to be as low as previously assumed" (referring to Arp's estimate).

The first serious attempt to measure the nova rate in M33 didn't come until the late 1980s when Della Valle analyzed a new batch of M33 plates taken between 1975 and 1986 at the Asiago Observatory, this time with the larger 182 cm telescope. Following the "control time method" used by Zwicky to estimate the frequency of supernovae in synoptic surveys (Zwicky 1942), Della Valle computed a nova rate of 4 ± 2 novae per year in M33. Over the next several years the nova rate in M33 was re-evaluated with estimates ranging from a low of just 0.45 yr^{-1} (Sharov 1993) to a high of 4.6 ± 0.9 yr^{-1} (Della Valle et al. 1994).

The most recent estimate of the nova rate in M33 was based on an Hα survey undertaken by Williams and Shafter using the 40 inch telescope at the Mount Laguna Observatory (Williams & Shafter 2004). The survey, which included observations made over eight epochs between 1995 and 2002, yielded a total of six novae. After correcting for the temporal and spatial completeness of the survey, a nova rate of $2.5^{+1.0}_{-0.7}$ yr^{-1} was estimated from numerical simulations that made use of the Hα light curves from Shafter and Irby's M31 survey (Shafter & Irby 2001). The M33 nova rate was also estimated using the mean nova lifetime methodology, which yielded a rate consistent with that found in the numerical simulations.

As in the case of M31, the published 2MASS magnitude ($K = 4.1$) is unreliable, so the $LSNR_K$ must be determined from optical observations coupled with a suitable optical-infrared color. Williams and Shafter found an M33 $LSNR_K$ of $2.34 \pm 0.99 \times 10^{-10}$ $L_{\odot,K}$ yr^{-1} based on an extinction-corrected V magnitude for M33 of $V_0 = 5.28$ coupled with a V–K color of 2.4 (Aaronson 1978).

Nova Populations in M33

M33, being a nearby low-mass, late-type galaxy without a significant bulge, is an important galaxy in the study of nova populations. Population synthesis models by Yungelson and collaborators in the late 1990s (Yungelson et al. 1997) predicted that the LSNRs should vary systematically across the Hubble sequence with low-mass, late-type galaxies having significantly higher LSNRs compared with more massive early-type, Population-II-dominated elliptical galaxies (see Section 7.5). A few years later, Matteucci et al. modeled the nova rates across a rage of Hubble type galaxies and came to similar conclusions (Matteucci et al. 2003). Given these predictions, it is surprising to discover that the $LSNR_K$ of M33 is comparable to, and perhaps even smaller than that of the bulge-dominated nova population in M31. Taken at face value these results are at variance with the predictions of population synthesis models that predict low-mass, late-type galaxies like M33 should be prolific nova producers.

Of the 48 nova candidates that have been detected in M33, a total of 15 recent systems have spectroscopic classifications. Specifically, there are eight Fe II, three Fe IIb, and four He/N class novae. The relatively high percentage of Fe IIb and He/N novae (near 50%) appears significantly larger than that seen in M31 and the Galaxy where the fraction is of order 20%. At the end of 2012, when only eight M33 novae had yet been spectroscopic classified, only two and possibly a third were classified as Fe II, with the other five being members of the He/N or Fe IIb classes. Based on this sample Shafter et al. (2012) estimated that the mix of spectroscopic classes differed from that seen in M31 and the Galaxy at the 99% confidence level. Although the numbers have been equaling out in recent years, the available evidence suggests that the fraction of Fe IIB and He/N novae is higher in later Hubble type galaxies such as the LMC and M33.

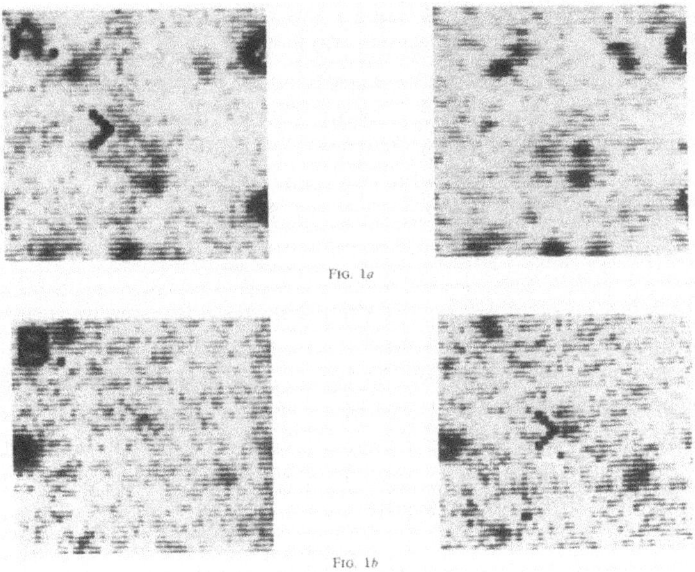

Figure 7.5. Images of the first two novae discovered in M87. At the time these were likely the faintest variable stars ever identified. (Reproduced from Pritchet and van den Bergh 1985. © 1985. The American Astronomical Society. All rights reserved.)

7.3 Beyond the Local Group

The increased sensitivity afforded by CCD detectors allowed for the study of novae beyond the Local Group. At about the same time as the Ciardullo et al. study was underway in M31, Chris Pritchet and Sidney van den Bergh were pushing nova surveys to their limit by imaging Virgo cluster ellipticals (Pritchet and van den Bergh 1987). Taking advantage of the large aperture and superb image quality available from the 3.6 m Canada–France–Hawaii Telescope (CFHT) perched atop the 14,000 ft peak of Mauna Kea in Hawaii, Pritchet and van den Bergh discovered two novae in a $2.2' \times 3.3'$ field in the Eastern part of M87 using an RCA 320×512 pixel CCD (see Figure 7.5). Nova #1 was seen on the decline, reaching apparent magnitude $B \leqslant 24.9$ on 1984 February 24, while the second nova was discovered on the rise, reaching $B \simeq 24.1$ four days later, on or just after February 28. As noted by Pritchet and van den Bergh in their paper, these were believed to be the faintest variable stars ever recorded.

Spurred by their successful detection of novae in M87, Pritchet and van den Bergh conducted a more systematic survey for novae in Virgo ellipticals during the spring of 1986. This time they avoided M87 because the high density of globular clusters near the core of the Galaxy complicated the discovery of faint variable sources. Instead, they focused on the first-ranked Virgo elliptical NGC 4472 (M49), and two other galaxies, NGC 4365 and NGC 4649. During 15 nights in March and April, they discovered a total of nine novae, with eight of these being detected in M49. After correcting for temporal coverage, Pritchet and van den Bergh estimated a combined nova rate of 160 ± 57 yr^{-1} in the surveyed fields of three Virgo cluster elliptical galaxies.

Pritchet and van den Bergh's Virgo survey was truly pioneering in a couple of noteworthy respects. It represented the first time that the specific nova rate was determined for any elliptical galaxies. By comparing the H-band luminosity-specific nova rate of the bulge of M31 with that for the three Virgo ellipticals, Pritchet and van den Bergh were able to show that the rates are consistent within the errors of the determinations. And also, through the use of the MMRD relation, Pritchet and van den Bergh were first to use novae to get distances to galaxies beyond the local group, finding a distance of 19.5 ± 3.9 Mpc to the Virgo cluster. Adopting a mean recession velocity, $\langle V \rangle = 1151 \pm 38$ km s^{-1} for 362 Virgo cluster galaxies (Hucra 1985), and correcting for the Virgo infall velocity and the solar motion, they estimated the Hubble constant to be 69 ± 14 km s^{-1} Mpc^{-1}, which turned out to be remarkably similar to the currently accepted value of $H = 72.8 \pm 3.1$ km s^{-1} Mpc^{-1} (Riess et al. 2016).

The early extragalactic nova surveys of Pritchet & van den Bergh (1987), and later Ciardullo et al. (1990a), ignited a broader interest in the stellar population of novae that has persisted ever since.

7.4 Novae in Globular Clusters

It has long been recognized that the number of (low-mass) X-ray binaries per unit mass is enhanced in Galactic globular clusters (GCs) relative to the field (e.g., Clark 1975; Katz 1975). A similar enhancement has long been recognized in M31 (e.g., Crampton et al. 1984; Trudolyubov & Priedhorsky 2004; Vulic et al. 2014). Given that the X-ray binaries are formed through stellar interactions involving neutron stars in the dense cores of GCs, it has been assumed that red dwarf–white dwarf binaries, including novae, should be similarly enhanced (e.g., Di Stefano & Rappaport 1994; Ivanova et al. 2006; Rivera-Sandoval et al. 2018).

Despite predictions that novae may be common in GCs, surprisingly few have been reported. Two targeted surveys covering the M31 GC system were undertaken in the 1990s. In the first, Ciardullo et al. (1990b) measured the brightnesses of 54 GCs that fell within their earlier M31 nova survey fields (Ciardullo et al. 1987) finding no increases that would be indicative of a nova eruption. Based on the effective survey time of $\sim$2 yr, they concluded that the overdensity of nova binaries was at least several times less than that for X-ray sources. Shortly afterward, (Tomaney et al. 1992) reported the results of a search for Hα emission in over 200 M31 GCs using a fiber-fed multi-object spectrograph. Since no other emission-line sources are expected in GCs, detection of Hα emission would signal the presence of a nova. Over an effective survey time of $\sim$1 yr for the entire M31 GC system, no nova eruptions were detected.

To date just seven novae have been found to erupt within GCs, with another three possibly associated with GCs. At least one nova, and possibly two more have been associated with Galactic GCs: T Sco (Nova Sco 1860) has been clearly identified with M80 (Luther 1860; Pogson 1860), whereas Nova Oph 1938 and a historical "guest star" in Chinese records from 58 BC may be associated with M14 and M22, respectively (Hogg 1964; Margon et al. 1991; Göttgens et al. 2019). Our neighboring galaxy M31 has produced the most definitive GC novae with three and possibly four so far identified: M31N 2007-06b in Bol 111 (Shafter & Quimby 2007), CXO J004345 in Bol 194 (Henze et al. 2009), M31N 2010-10f in Bol 126 (Cao et al. 2012; Henze et al. 2013), and possibly an anonymous source in Bol 383 (Peacock et al. 2010). M87, with its high nova rate and large GC population is next with two GC novae so far identified. The first was discovered by Shara et al. (2004) with the other recently identified by Curtin et al. (2015) in their ground-based Virgo nova survey. In the same survey, Curtin et al. found another nova to be coincident with a GC in M84.

It seems likely that many novae have been missed in GCs as a consequence of the added difficulty in discovering novae against the background light of the cluster. Curtin et al. studied the enhancement of novae in GCs relative to the field in the three Virgo ellipticals included in their survey estimating that novae erupt 10 times more frequently in GCs than they do in the field. Specifically, they estimated that novae occur in M87 GCs with a frequency, $f \simeq 4 \times 10^{-4}$ novae per GC per year. This is an order of magnitude less than the frequency of $f \simeq 4 \times 10^{-3}$ that was estimated earlier by Shara et al.

The number of novae discovered thus far in GCs is too limited to establish whether there are any systematic differences in their behavior relative to field novae. Theoretical models of nova light curves from low-metallicity Population II environments typical of GCs have been performed by Kato et al. (2013) who predict that novae from the lowest metallicity GCs should be relatively faint and slow. In such clusters only novae occurring on the highest mass white dwarfs should be easily detectable. If so, optical surveys are likely missing a significant fraction of novae erupting in GCs because they are systematically fainter than their Population I counterparts. Kato et al. stress the importance of X-ray observations as an important avenue for the detection of novae in GCs in future surveys.

7.5 Nova Rates Across the Hubble Sequence

By the early 1990s nova rates had been estimated in a number of galaxies spanning the full range of Hubble types making it possible to explore whether the observed properties of novae (e.g., their peak luminosities, rates of decline from maximum light, and rates of occurrence) varied with the underlying stellar population of a given galaxy.

The first such study was published by Ciardullo et al. as part of their nova survey of the elliptical component of NGC 5128 (Ciardullo et al. 1990a). At the time of their study, nova rates had been estimated for a total of just five galaxies: M31, M33, LMC, SMC, and NGC 5128. After estimating the infrared K-band luminosities for each galaxy, Ciardullo et al. determined the $LSNR_K$ values shown in Figure 7.6. In addition, they were able to add the $LSNR_K$ for the surveyed fields of the three Virgo cluster elliptical galaxies studied by Pritchet & van den Bergh. Ciardullo et al. found no evidence that the $LSNR_K$ varied systematically with Hubble type, although the uncertainties were large, and the number of galaxies was limited.

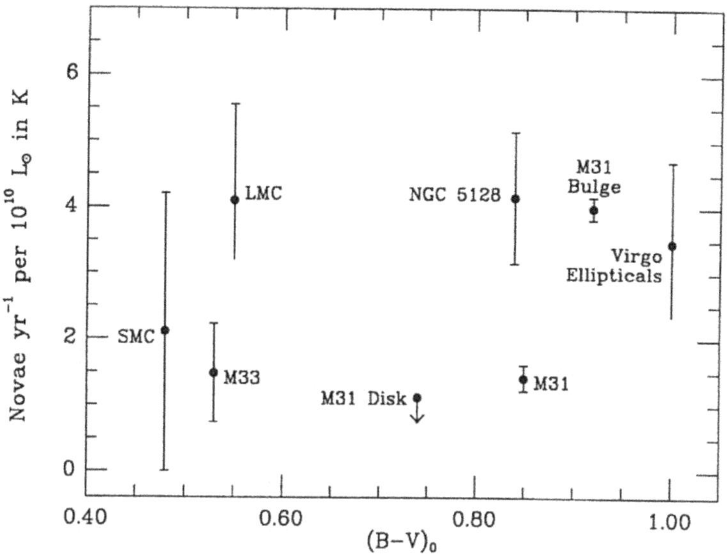

Figure 7.6. A comparison of the LSNRs from the six galaxies where nova rates had been determined. For M31, estimates of the bulge and disk components are shown separately. The Virgo ellipticals point corresponds to the $LSNR_K$ for Virgo elliptical galaxies studied by Pritchet and van den Bergh (1987). (Reproduced from Ciardullo et al. 1990a.)

Within a few years after the Ciardullo et al. study appeared, another analysis of LSNRs across galaxy type was published by Della Valle and collaborators. Their work was prompted by a redetermination of the nova rate in M33, the addition of data for M81, and a re-analysis of the infrared luminosities of the host galaxies. In particular, Della Valle et al. (1994) argued that the luminosity of the surveyed regions of NGC 5128 and the Virgo ellipticals from the Pritchet and van den Bergh study had been underestimated. In the former case, because of underestimating the extinction, and in the latter case, by underestimating the distance to the Virgo cluster. The Della Valle et al. analysis yielded the LSNRs shown in Figure 7.7, suggesting, contrary to the Ciardullo et al. study, that bulge-less, late-type galaxies such as the LMC and M33 were more prolific nova producers than the earlier type bulge-dominated galaxies.

Shortly after the Della Valle et al. study appeared, the first serious population synthesis study addressing the issue of nova rates in different Hubble type galaxies was published by Yungelson et al. (1997). Based on population synthesis models for newly formed cataclysmic variable systems, which showed that the average mass of the white dwarf primaries decreased with increasing time having elapsed from the formation of the zero age main sequence, Yungelson et al. realized that the nova rate and other properties should depend on a galaxy's star formation history. As expected, their models demonstrated that galaxies that have experienced more recent star formation, such as irregular and low-mass, disk-dominated galaxies, produce more novae for a given *K*-band luminosity than do galaxies that experienced a higher rate of early star formation, such as massive spirals and elliptical

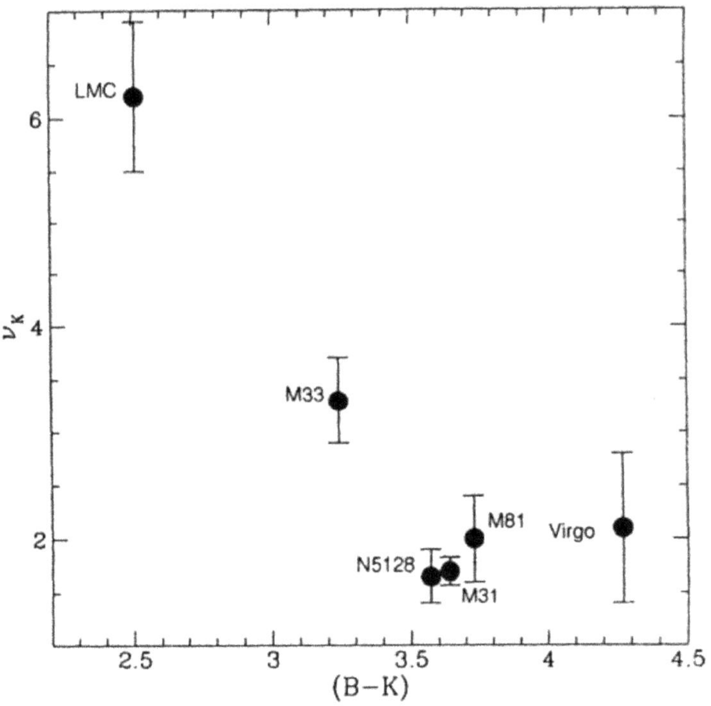

Figure 7.7. The infrared *H*-band LSNR of galaxies as a function of color. (Reproduced with permission from Della Valle et al. 1994, © ESO.)

galaxies. Their results were in broad agreement with the findings of Della Valle's group.

Over the next two decades the predictions of the population synthesis models were put to the test by observations of novae in additional galaxies. At the dawn of the new millennium, Shafter, Ciardullo, and Pritchet published a nova survey conducted with the Kitt Peak National Observatory's signature 4 m reflector designed to measure and compare the LSNRs in the spiral galaxies M51 and M101 with that of the Virgo elliptical, M87. During four epochs over four consecutive years (1994–1997), a total of nine and 12 novae were discovered in the spirals M51 and M101, respectively. The Virgo elliptical M87 was only observed over two epochs, with nine novae identified. After correcting for temporal sampling, the mean nova lifetime method was employed to estimate nova rates of 18 ± 7, 91 ± 34, and 12 ± 4 per year for M51, M87 and M101, respectively, which correspond to luminosity-specific nova rates of 1.09 ± 0.47, 2.30 ± 0.99, and 0.97 ± 0.38 novae per year per 10^{10} solar luminosities in K. To the extent that the differences in the LSNRs are significant, the spiral systems appear to have *lower*, rather than higher LSNRs as predicted by the population synthesis models.

A major issue that has always plagued the determination of extragalactic nova rates concerns the estimation of limiting magnitudes and survey completenesses in the bright central regions of galaxies. As we have seen, the early broad-band photometric surveys of M31 by Hubble and Arp were essentially blind to novae occurring within ~4′ of the nucleus. The problem is even worse in the case of elliptical galaxies with their extremely high central brightness. An obvious way to mitigate the difficulty of detecting novae against the bright stellar background is to employ space-based imaging where the point-spread function of the stellar images are small, and the background light can be greatly minimized. With these considerations in mind, in 2003 Laura Ferrarese, Patrick Côte, and Andrés Jordán published the first HST study of extragalactic novae—their target being the first-ranked Virgo elliptical galaxy, M49 (Ferrarese et al. 2003).

The Ferrarese et al. study produced a number of surprising results. The targeted *HST* survey covered 19 epochs spanning a period of 55 days and yielded a total of nine novae. The relatively frequent sampling produced light curves that were sufficiently well sampled to determine peak magnitudes and fade rates for most of the novae. Figure 7.8 shows the distribution of fade rates for the M49 novae compared with novae in M31, the LMC and the Galaxy. Surprisingly, the M49 novae were generally *faster* than their counterparts in the later type galaxies—opposite to the predictions of the population synthesis models.

After correcting for completeness, a Monte Carlo numerical analysis led to an estimated nova rate of ~100 yr^{-1}, and a LSNR_K of $1.71 \pm 0.61\ \mathrm{yr}^{-1}\ 10^{-10}\ \mathrm{L}_{K,\odot}$ ($\nu_K = 2.52 \pm 0.91\ \mathrm{yr}^{-1}\ 10^{-10}\ \mathrm{L}_{K,\odot}$ when the less reliable 2MASS photometry was used). These results did not suggest that nova rates had been systematically underestimated in ground-based surveys, as feared. Rather, the LSNR_K was typical of other galaxies.

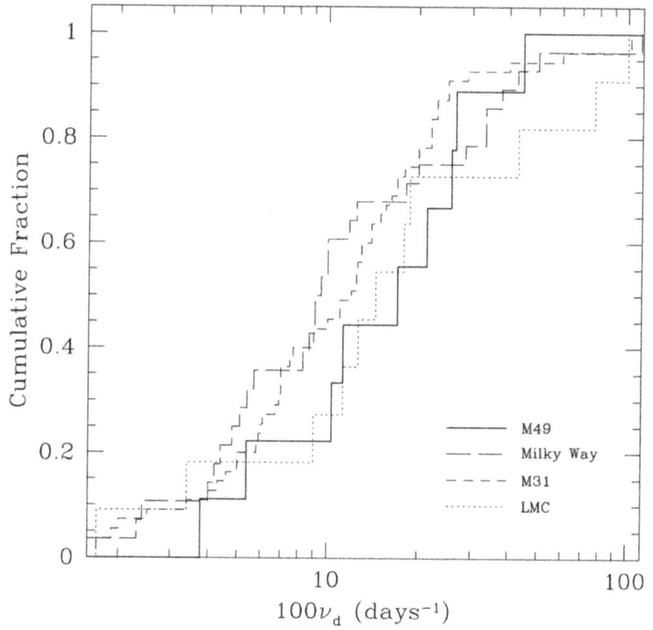

Figure 7.8. The cumulative distribution of fade rates for M49, the Galaxy, M31, and the LMC. Note that the M49 novae (like those in the LMC) are generally "faster" than their counterparts in M31 and the Galaxy. (Reproduced from Ferrarese et al. 2003. © 2003. The American Astronomical Society. All rights reserved.)

At about the same time that Ferrarese et al. were studying M49, Michael Shara and collaborators were studying *HST* archival images of another Virgo cluster giant elliptical, M87. Preliminary results of their study were published in the proceedings of a nova conference held in Sitges, Spain in 2002 May. Although a formal determination of the nova rate was not attempted, the authors concluded that the nova rate in M87 is "300 events per year or more." The discrepancy between these results and those of Ferrarese et al. for M49 were puzzling. Both studies were based on *HST* observations of galaxies with very similar Hubble types and distances.

The situation became even more confusing in 2007 when Juan Madrid and collaborators reported the discovery of 13 near-UV transients that they identified as nova candidates near the core of M87. Remarkably, 11 out of the 13 sources were found within 9″ of the nucleus. If indeed these sources are novae, an extrapolation of the surveyed region to the entire galaxy, would suggest a total nova rate of more than 1000 per year! Given that such a high nova rate seemed implausible, Madrid et al. concluded that the novae in M87 did not follow the overall galactic light, but were highly concentrated toward the central regions of the Galaxy.

To summarize, after the publication of the Madrid et al. study, it was fair to conclude that our knowledge of the nova rates in M87 and M49 were quite uncertain, with estimates in M87 ranging from 100 to 300 per year, or perhaps even higher. By 2010, it seemed clear that our understanding of whether the LSNRs vary across the Hubble sequence would require a more determined effort to accurately measure the nova rate in the Virgo ellipticals. This uncertainty, coupled with the possibility that M87 might have a significantly higher $LSNR_K$ compared to its slightly more luminous Virgo Cluster companion, M49, led Shafter and collaborators to consider whether the difference in the globular cluster populations of the two galaxies might play a role in their differing nova rates. The unusually large globular population of M87 is well known. With a specific frequency, $S_N \sim 14$, M87 has approximately three times the number of globular clusters as M49.[2] Was it just a coincidence that their nova rates appeared to also differ by roughly a factor of three? In order to shed light on this question, Shafter and Pritchet proposed to conduct a survey of the Virgo ellipticals using the MegaCam on the 3.6 m Canada–France–Hawaii telescope (CFHT). The CHFT is the perfect ground-based telescope for this study, given the spectacular seeing that this telescope is renowned for.

Over the course of 14 months during 2011 and 2012, Shafter and collaborators acquired four epochs of observations of three Virgo elliptical galaxies, M49, M84, and M87, which have globular cluster specific frequencies of 14, 3.6, and 1.6, respectively. To minimize the calibration errors and simplify the analysis, all three galaxies were observed on the same nights, with the same integration times, and through the same (Hα) filter. Thus the relative number of novae discovered in the three galaxies should reflect their relative nova rates. At the conclusion of the survey a total of 27, 37, and 19 nova candidates were found in M87, M49, and M84, respectively, with no apparent dependence on the globular cluster specific frequency.

[2] The specific frequency, $S_N = N_{\rm GC}10^{0.4(M_V+15)}$, is the number of globular clusters, $N_{\rm GC}$, per unit galaxy luminosity, normalized to a galaxy with $M_V = -15$ (Harris and van den Bergh 1981).

In addition to the galaxy-to-galaxy comparison, nova rates for the individual galaxies were estimated using numerical simulations. For M87, M49, and M84, respectively, rates of 154^{+23}_{-19}, 189^{+26}_{-22}, and 95^{+15}_{-14} were found, corresponding to specific rates of $\nu_K = 3.8 \pm 1.0$, 3.4 ± 0.6, and 3.0 ± 0.6 novae per year per 10^{10} solar luminosities in the *K* band. The value for M87 was about 50% higher that that found in the previous ground-based survey conducted with the KPNO 4 m (Shafter et al. 2000), while the M49 estimate was nearly a factor of two higher than that found a decade earlier by Ferrarese et al. using *HST* observations.

The CFHT Virgo study would not be the final word on the nova rate in M87, however. In the past few years, two important M87 nova rate determinations have been published, both of which were based on *HST* archival data from a microlensing survey undertaken in *HST* Cycle 14 by E. A. Baltz (program 10543). The data, which were taken with the *HST* Advance Camera for Surveys (ACS), cover a 202″ × 202″ region centered on the nucleus of M87. A total of 61 epochs (visits) were obtained over a time span of 72 days between 2005 December 24 and 2006 March 5. In the first analysis of the data, a group led by Michael Shara identified a total of 32 classical novae. After correcting for temporal sampling and the fraction of M87's light covered in the survey, Shara et al. estimated a nova rate of 363^{+33}_{-45} yr^{-1}, corresponding to a LSNR_K of $\nu_K = 7.88^{+2.3}_{-2.6}$ yr^{-1} $[10^{10}\ L_{K,\odot}]^{-1}$.

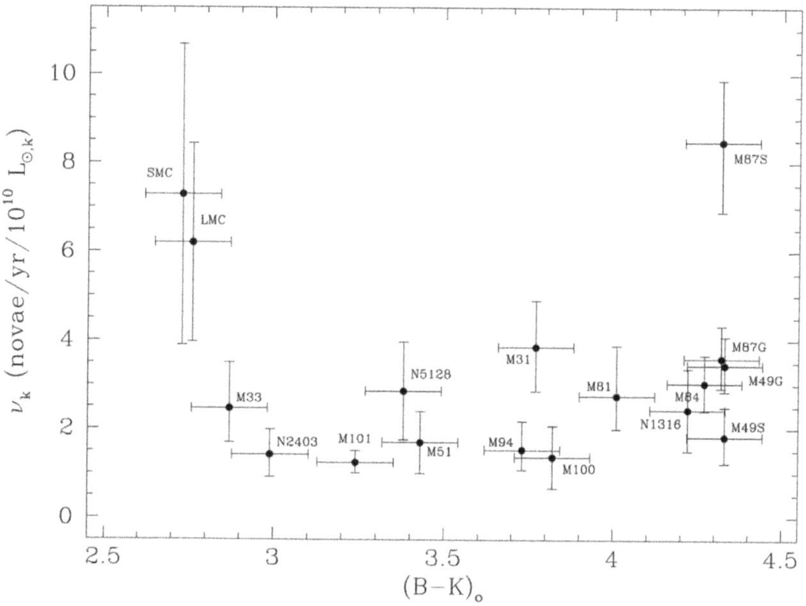

Figure 7.9. A comparison of the most current estimates of the LSNRs for galaxies with measured nova rates. The M49G (and M87G) and M49S (and M87S) designations refer to ground- and space-based LSNR_K determinations, respectively. There is no evidence that the LSNR_K varies systematically across the Hubble sequence of galaxy types. It is possible that the mean LSNR_K for galaxies other than the Magellanic Clouds and M87 have been systematically underestimated due to low cadence surveys that may have missed the faintest and fastest evolving novae.

Table 7.3. Galaxies with Measured Luminosity-specific Nova Rates

Galaxy	Hubble type	B–K	Nova Rate (yr^{-1})	LSNR_K (yr^{-1} $[10^{10}\ L_{K,\odot}]^{-1}$)	Reference
SMC	SB(s)m pec	2.73	0.9 ± 0.4	7.5 ± 3.1	Mróz (2016)
LMC	SB(s)m	2.76	2.4 ± 0.8	6.5 ± 2.2	Mróz (2016)
M31	SA(s)b	3.77	65^{+16}_{-15}	3.85 ± 1.0	Darnley et al. (2006)
M33	SA(s)cd	2.87	$2.5^{+1.0}_{-0.7}$	$2.45^{+1.0}_{-0.8}$	Williams & Shafter (2004)
M49 Ground	E2	4.33	189^{+26}_{-22}	$3.44^{+0.64}_{-0.59}$	Curtin et al. (2015)
M49 Space	E2	4.33	100^{+35}_{-30}	$1.82^{+0.68}_{-0.59}$	Ferrarese et al. (2003)
M51	SA(s)bc pec	3.43	18 ± 7	1.69 ± 0.7	Shafter et al. (2000)
M81	SA(s)ab	4.01	33^{+13}_{-8}	$2.74^{+1.14}_{-0.75}$	Neill & Shara (2004)
M84	E1	4.27	95^{+15}_{-14}	$3.03^{+0.64}_{-0.61}$	Curtin et al. (2015)
M87 Ground	E + 0 − 1 pec	4.32	154^{+23}_{-19}	$3.59^{+0.73}_{-0.66}$	Curtin et al. (2015)
M87 Space	E + 0 − 1 pec	4.32	363^{+33}_{-45}	$7.9^{+2.3}_{-2.6}$	Shara et al. (2016)
M94	SA(r)ab	3.73	$5.9^{+2.4}_{-1.6}$	$1.52^{+0.64}_{-0.45}$	Güth (2010)
M100	SAB(s)bc	3.82	25 ± 12.5	1.36 ± 0.70	Ferrarese et al. (1996)
M101	SAB(rs)cd	3.24	$11.7^{+1.9}_{-1.5}$	$1.24^{+0.27}_{-0.24}$	Coelho et al. (2008)
NGC 1316	SAB(s)0^0	4.22	135 ± 45	2.43 ± 0.93	Della Valle & Gilmozzi (2002)
NGC 2403	SAB(s)cd	2.99	$2^{+0.5}_{-0.3}$	$1.41^{+0.56}_{-0.51}$	Franck et al. (2012)
NGC 5128	S0 pec	3.38	8 ± 2.8	2.85 ± 1.10	Ciardullo et al. (1990a)

An independent analysis of the archival *HST* data was undertaken by Shafter and collaborators who found results that were broadly consistent with that found by Shara's group. The analysis yielded 33 transient sources, with most, but not all, independently identified in the Shara et al. study. The global nova rate for M87 depends on three factors or assumptions, all of which are relatively uncertain: (1) the fraction of the transient sources that are bona fide classical novae, (2) the assumption that the specific nova rate is constant with distance from the center of M87, and (3) the size of M87's extended halo. If all sources are considered novae, with specific rates assumed to be independent of spatial position in M87, a nova rate of $339^{+37}_{-33}\ \mathrm{yr}^{-1}$ is found, in good agreement with the Shara et al. result. However, when only transients with high quality light curves that confirm their identification as novae are considered, the estimated nova rate drops to $215^{+35}_{-27}\ \mathrm{yr}^{-1}$. In addition, as the Madrid et al. survey showed, the nova rate in M87 appears to be significantly higher in the inner regions of the galaxy. If so, the extrapolation of the nova rate in the surveyed region to the entire galaxy by assuming a uniform specific nova rate will overestimate the nova rate for the galaxy as a whole. Given these considerations, Shafter et al. argue that a rate of $\sim$350 yr^{-1} is best considered an upper limit to the true nova rate in M87.

A summary of our present understanding of how the LSNRs vary across the Hubble sequence is presented in Figure 7.9. The figure incorporates the best $LSNR_K$ estimates currently available for all galaxies where nova rates have been measured (see Table 7.3). Based on the current data there is little evidence to support the idea that the LSNRs vary systematically with galaxy Hubble type, and hence the dominant progenitor stellar population. Indeed, the highest measured LSNRs are found at opposite ends of the Hubble sequence with values for the Magellanic clouds consistent with that found for the Virgo elliptical galaxy M87. It is interesting that these relatively high LSNRs represent galaxies with the most recently determined values. The Magellanic Cloud studies were ground-based, and a product of the OGLE program, while the M87 study was based on data from the *HST*. Despite their differences, both studies had the advantage of a relatively high cadence, which assured that the surveys would be sensitive to rapidly-evolving novae. It is tempting to speculate that the LSNRs for the majority of galaxies may have been systematically underestimated in prior studies, and that a galaxy's $LSNR_K$ in actuality is insensitive to Hubble type and is characterized by a value of $\nu_K = \sim 7$ yr^{-1} $[10^{10}\ L_{K,\odot}]^{-1}$. This possibility will be explored further in Chapter 8.

References

Aaronson, M. 1978, ApJ, 221, 103

Arp, H. C. 1956, AJ, 61, 15

Brimacombe, J., Vallely, P., & Stanek, K. Z. 2018, ATel #11470

Cao, Y., Kasliwal, M., & Neill, J. D. 2012, ApJ, 752, 133

Capaccioli, M., Della Valle, M., D'Onofrio, M., et al. 1989, AJ, 97, 1622

Capaccioli, M., Della Valle, M., D'Onofrio, M., & Rosino, L. 1990, ApJ, 360, 63

Chen, H.-L., Woods, T. E., Yungelson, L. R., Gilfanov, M., & Han, Z. 2016, MNRAS, 458, 2916

Ciardullo, R., Ford, H. C., Neill, J. D., et al. 1987, ApJ, 318, 520

Ciardullo, R., Ford, H. C., Williams, R. E., et al. 1990a, AJ, 99, 1079

Ciardullo, R., Tamblyn, P., & Phillips, A. C. 1990b, PASP, 102, 1113

Clark, G. W. 1975, ApJ, 199, L143

Coelho, E. A., Shafter, A. W., & Misselt, K. A. 2008, ApJ, 686, 1261

Crampton, D., Hutchings, J. B., Cowley, A. P., et al. 1984, ApJ, 284, 663

Crayton, H., Rector, T. A., Walentosky, M. J., et al. 2018, AAS, 232, 320.08

Curtin, C., Shafter, A. W., Pritchet, C. J., et al. 2015, ApJ, 811, 34

Darnley, M. J., Bode, M. F., Kerins, E., et al. 2004, MNRAS, 353, 571

Darnley, M. J., Bode, M. F., Kerins, E., et al. 2006, MNRAS, 369, 257

de Vaucouleurs, G., de Vaucouleurs, A., Corwin, H. G., Buta, R. J., Paturel, G., & Fouque, P. 1991, Third Reference Catalogue of Bright Galaxies (RC3) (New York: Springer)

Della Valle, M., & Gilmozzi, R. 2002, Sci., 296, 1275

Della Valle, M., & Livio, M. 1998, ApJ, 506, 818

Della Valle, M., Rosino, L., Bianchini, A., & Livio, M. 1994, A&A, 287, 403

Di Stefano, R., & Rappaport, S. 1994, ApJ, 423, 274

Ferrarese, L., Côté, P., & Jordán, A. 2003, ApJ, 599, 1302
Ferrarese, L., Livio, M., Freedman, W., et al. 1996, ApJL, 468, 95
Franck, J. R., Shafter, A. W., Hornoch, K., & Misselt, K. A. 2012, ApJ, 760, 13
Freedman, W. L., Madore, B. F., Gibson, B. K., et al. 2001, ApJ, 553, 47
Göttgens, F., Weilbacher, P. M., Roth, M. M., et al. 2019, arXiv: 1904.11515
Graham, J. A. 1979, in Proc. IAU Coll. 46, Changing Trends in Variable Star Research, ed. F. Bateson, J. Smak, & I. H. Urch (Hamilton: University of Waikato), 96
Güth, T., Shafter, A. W., & Misselt, K. A. 2010, ApJ, 720, 1155
Harris, W. E., & van den Bergh, S. 1981, AJ, 86, 1627
Hatano, K., Branch, D., Fisher, A., & Starrfield, S. 1997, ApJ, 487, 45
Healy, M. W., Darnley, M. J., Copperwheat, C. M., et al. 2019, MNRAS, 486, 4334
Henze, M., Pietsch, W., Haberl, F., et al. 2009, A&A, 500, 769
Henze, M., Pietsch, W., Haberl, F., et al. 2013, A&A, 549, 120
Hogg, D. E. 1964, JRASC, 58, 203
Hubble, E. 1926, ApJ, 63, 236
Hubble, E. 1929, ApJ, 69, 103
Ivanova, N., Heinke, C. O., Rasio, F. A., et al. 2006, MNRAS, 372, 1043
Jarrett, T. H., Chester, T., Cutri, R., et al. 2003, AJ, 125, 525
Kato, M., Hachisu, I., & Henze, M. 2013, ApJ, 779, 19
Katz, J. I. 1975, Natur, 253, 698
King, J. Y., & Li, W. D. 1999, IAU Circ., 7208
Luther, E. 1860, AN, 53, 293
Luyten, W. J. 1927, BHarO, 847, 8
Mantegazza, L., Antonello, E., Fugazza, D., et al. 2001, A&A, 367, 759
Margon, B., Anderson, S. F., Downes, R. A., et al. 1991, ApJL, 369, L71
Matteucci, F., Renda, A., Pipino, A., & Della Valle, M. 2003, A&A, 405, 23
McKibbem, V. N. 1951, BHarO, 920, 13
Mróz, P., Udalski, A., Poleski, R., et al. 2016, ApJS, 222, 9
Neill, J. D., & Shara, M. M. 2004, AJ, 127, 816
Özdönmez, A., Güver, T., Cabrera-Lavers, A., & Ak, T. 2018, MNRAS, 461, 1177
Peacock, M. B., Maccarone, T. J., Kundu, A., & Zepf, S. E. 2010, MNRAS, 407, 2611
Pietsch, W. 2010, AN, 331, 187
Pogson, N. 1860, MNRAS, 21, 32
Pritchet, C. J., & van den Bergh, S. 1985, ApJ, 288, L41
Pritchet, C. J., & van den Bergh, S. 1987, ApJ, 318, 507
Riess, A. G., Macri, Lucas, M., Hoffmann, S. L., et al. 2016, ApJ, 826, 56
Rivera-Sandoval, L. E., van den Berg, M., Heinke, C. O., et al. 2018, MNRAS, 475, 484
Rosino, L. 1964, AnAp, 27, 498
Rosino, L. 1973, A&AS, 9, 347
Rosino, L., & Bianchini, A. 1973, A&A, 22, 461
Rosino, L., Capaccioli, M., D'Onofrio, M., & Della Valle, M. 1989, AJ, 97, 83
Sandage, A. 1971, ApJ, 166, 13
Sandage, A. R., Becklin, E. E., & Neugebauer, G. 1969, ApJ, 157, 55
Schlegel, D. J., Finkbeiner, D. P., & Davis, M. 1998, ApJ, 500, 525
Shafter, A. W. 2013, AJ, 145, 117
Shafter, A. W. 2017, ApJ, 834, 196

Shafter, A. W., Ciardullo, R., & Pritchet, C. J. 2000, ApJ, 530, 193
Shafter, A. W., Darnley, M. J., Bode, M. F., & Ciardullo, R. 2012, ApJ, 752, 156
Shafter, A. W., Darnley, M. J., Hornoch, K., et al. 2011, ApJ, 734, 12
Shafter, A. W., & Irby, B. K. 2001, ApJ, 563, 749
Shafter, A. W., & Quimby, R. M. 2007, ApJL, 671, L121
Shara, M. M., Doyle, T. F., Lauer, T. R., et al. 2016, ApJS, 227, 1
Shara, M. M., Zurek, D. R., Baltz, E. A., et al. 2004, ApJL, 605, L117
Sharov, A. S. 1993, AstL, 19, 147
Sharov, A. S., & Alksnis, A. 1991, Ap&SS, 180, 273
Tomaney, A., Crotts, A., & Shafter, A. 1992, AAS, 181, 73.09
Trudolyubov, S., & Priedhorsky, W. 2004, ApJ, 616, 821
Udalski, A., Kubiak, M., & Szymański, M. K. 1997, AcA, 47, 319
Udalski, A., Szymański, M. K., Soszyński, I., & Poleski, R. 2008, AcA, 58, 69
Udalski, A., Szymański, M. K., & Szymański, G. 2015, AcA, 65, 1
Vulic, N., Gallagher, S. C., & Barmby, P. 2014, ApJ, 790, 136
Wei, J. Y., Xu, D. W., Qiao, Q. Y., et al. 1999, IAU Circ., 7209
Williams, R. E. 1992, AJ, 104, 725
Williams, S. C., Darnley, M. J., Bode, M. F., et al. 2014b, ApJS, 213, 10
Williams, S. C., Darnley, M. J., Bode, M. F., & Shafter, A. W. 2014a, in ASP Conf. Ser. 490, Stella Novae: Past and Future Decades, ed. P. A. Woudt, & V. A. R. M. Ribeiro (San Francisco, CA: ASP), 85
Williams, S. C., Darnley, M. J., Bode, M. F., & Shafter, A. W. 2016, ApJ, 817, 143
Williams, S. C., Darnley, M. J., & Henze, M. 2015, MNRAS, 472, 1300
Williams, S. J., & Shafter, A. W. 2004, ApJ, 612, 867
Willmer, C. N. A. 2018, ApJS, 236, 47
Yungelson, L., Livio, M., & Tutukov, A. 1997, ApJ, 481, 127
Zwicky, F. 1942, ApJ, 96, 28

AAS | IOP Astronomy

Extragalactic Novae
A historical perspective
Allen W Shafter

Chapter 8

Outstanding Questions and Future Prospects

Despite the wealth of data produced by the surveys described earlier, our understanding of many fundamental properties of novae remains incomplete. From an observational perspective, something as basic as the luminosity function of novae (the distribution of luminosities at the peak of the eruption) remains poorly characterized. Similarly, the distribution of decline rates, and how they may relate to the peak luminosity (i.e., the MMRD relation), is also unresolved. In this chapter, I will review our current state of knowledge on these topics, and discuss what observations will be required to make significant progress in these areas. From the point of view of theory, there are also unresolved questions with regard to the nature of the thermonuclear runaway, and how it depends on properties of the progenitor binary. Although important, such considerations are beyond the scope of our discussion, which is primarily observational.

We begin by reviewing what has been learned about the luminosity function of novae, and how these properties may vary with the underlying stellar population. We will then revisit the MMRD relation in light of recent observations that have revealed a population of faint but fast novae that violate the MMRD. Many of these systems appear to be recurrent novae, such as M31N 2008-12a, which has a remarkably short recurrence time of only a year. Given that such systems are hard to detect observationally, a major unanswered question is just how common such systems actually are. Finally, we will conclude by describing future observations that will help to shed light on how extragalactic nova rates vary across the Hubble sequence.

8.1 The Luminosity Function of Novae

Nova eruptions are unpredictable, and historically it has been difficult to obtain the complete light curves necessary to determine with confidence the peak magnitudes and rates of decline. Given how fast some novae evolve, without daily monitoring it is impossible to be sure that the discovery magnitude corresponds to peak

doi:10.1088/2514-3433/ab2c63ch8

brightness. Achieving this sort of high temporal sampling has rarely been possible using telescopes scheduled in the traditional fashion with short blocks of time allocated monthly, and well in advance. The situation has improved considerably in recent years thanks to the more widespread use of queue telescope scheduling (where an observer has the potential of being granted a long series of partial nights, rather than a shorter period of full nights), and the proliferation of automated survey telescopes.

8.1.1 The Maximum Magnitude Distribution

Shafter et al. (2009) studied the absolute magnitude distributions of a large sample of M31 novae and compared it to similar distributions for the Galaxy and the LMC (see Figure 8.1). The M31 sample, which includes 732 nova candidates discovered up through 2007, were from the online database of M31 novae kept by Wolfgang Pietsch at the Max Planck Institute (Pietsch 2010). The Galactic data were taken from the compilations found in the work by Shafter (1997) and Downes & Duerbeck (2000), while the LMC data were computed from the data given in Shida & Liller (2004) assuming $\mu_0(\mathrm{LMC}) = 18.5$ and $A_V = 0.4$ (van den Bergh 2000). Although the Galactic, and especially the LMC samples are relatively small, the data suggest that the M31 novae may be fainter on average at maximum light ($\langle M_V[\mathrm{M31}]\rangle = -7.20 \pm 0.04$, $\langle M_V[\mathrm{MW}]\rangle = -7.79 \pm 0.11$, $\langle M_V[\mathrm{LMC}]\rangle = -8.06 \pm 0.05$). However, several selection effects are almost certainly playing a role here. For example, the Galactic novae are typically heavily extincted, making it more likely that a greater fraction of fainter novae may be missed. In the case of M31, which is surveyed less frequently, it is possible that fewer novae are discovered at maximum light. As a result, the M31

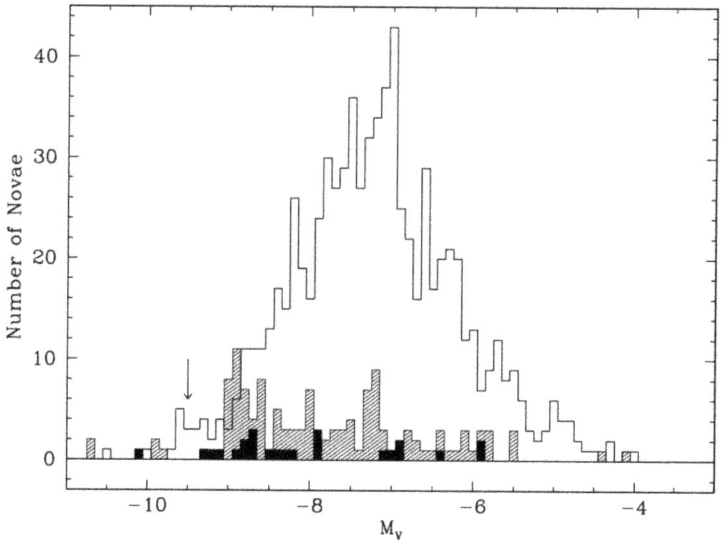

Figure 8.1. The distribution of absolute magnitudes at discovery for 732 novae in M31 compared with samples from the Milky Way (shaded region), and the LMC (black region). (Reproduced from Shafter et al. 2009. © 2009. The American Astronomical Society. All rights reserved.)

distribution should best be considered an upper limit to the true peak absolute magnitude distribution. The difference should not amount to more than a few tenths of a magnitude given that in most cases the novae did not fade significantly prior to discovery.

More recently, Shafter (2017) reconsidered the M31 distribution, again using the Pietsch database, and found the updated distribution shown in Figure 8.2(a). The distribution of absolute magnitudes of M31 novae is approximately Gaussian, with a peak magnitude and standard deviation of $M_V = -7.2$ and 0.8 mag, respectively. For comparison Figure 8.2(b) shows the absolute magnitude distribution for an updated sample of Galactic novae from the work of Aykut Özdönmez and collaborators (Özdönmez et al. 2018). Again, the mean absolute magnitude of the Galactic sample appears to be higher than that for M31. Whether or not this represents a real population difference (the M31 sample is mainly bulge novae, while the Galactic sample is dominated by a disk novae, mostly in the solar neighborhood), or can be explained by selection bias remains to be seen.

8.1.2 Rates of Decline

The measurement of nova decline rates are challenging both because the novae must be followed for several magnitudes past maximum light, but also because the declines are often not smooth, exhibiting a wide variety of morphologies. Richard Strope and Bradley Schaefer along with Arne Henden of the American Association of Variable Star Observers (AAVSO) were able to assemble a large sample of Galactic nova light curves drawing heavily from the vast AAVSO data archive (Strope et al. 2010). They settled on six representative light curve morphologies—Flat top (F), Cusp (C), Jitters (J), Plateau (P), Oscillations (O), Smooth (S), and Dust dip (D)—with the prototypes shown in Figure 8.3.

Clearly, the measurement of decline rates are complicated by the jiggles, oscillations, and dust-formation events exhibited by many classes of novae.

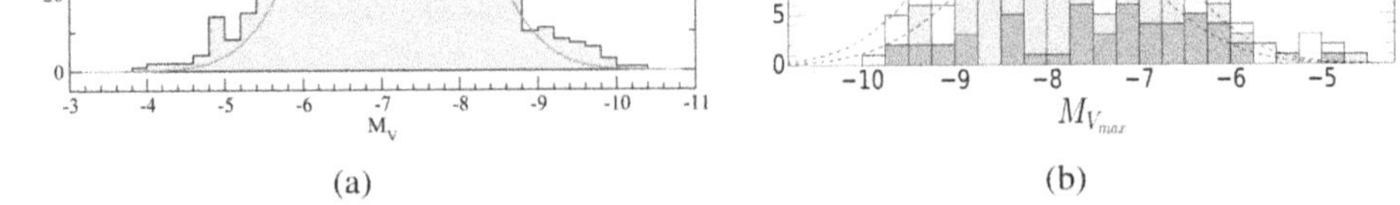

Figure 8.2. The absolute magnitude distribution of novae from recent works by Shafter (2017; left) and Özdönmez et al. (2018; right). Sample 1 refers to novae with the most reliable distances, whereas samples 2a and 2b rely on MMRD distances with reliable and unreliable light curve parameters, respectively. (Left: © 2017. The American Astronomical Society. All rights reserved. Right: Copyright of OUP © 2018.)

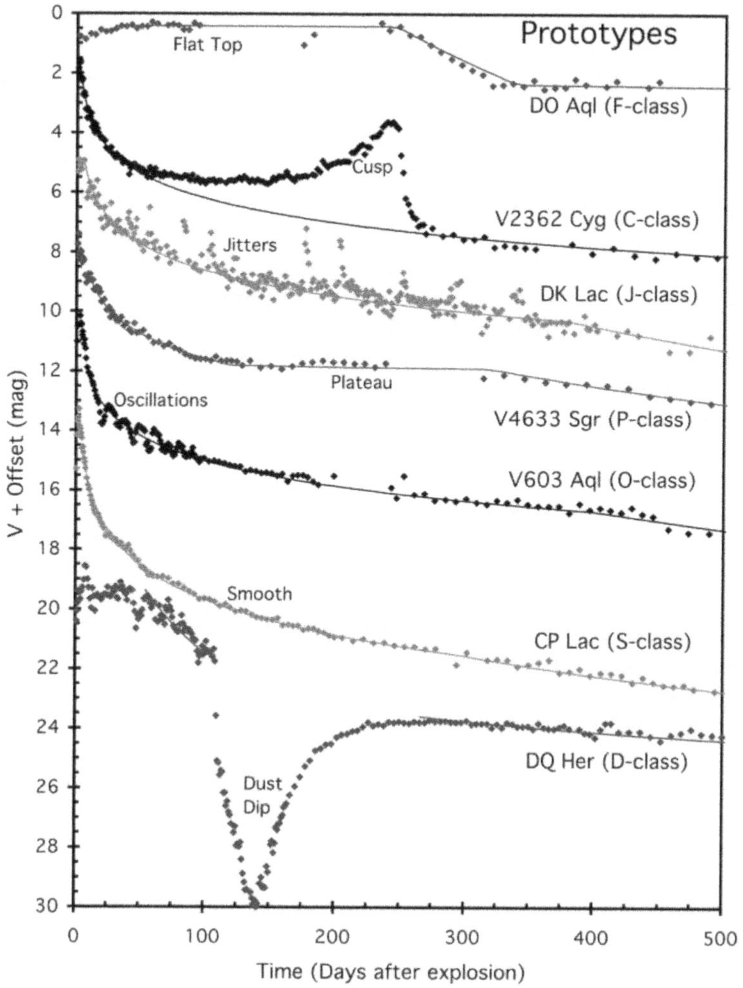

Figure 8.3. Examples of nova light curves in the morphological classification scheme of Strope et al. (2010). The diversity in light curve behavior makes efforts to determine fade rate parameters, t_2 and t_3, problematic. See text for details. (Reproduced from Strope et al. 2010. © 2010. The American Astronomical Society. All rights reserved.)

The longer from peak luminosity, the more likely it becomes that these features will affect the decline rate measurements. Thus, generally speaking, t_2 times should provide a more accurate and robust indicator of decline rate than will t_3 times, which can be significantly affected in dust forming novae. Figure 8.4 shows the distribution of t_3 times for a large sample of Galactic novae studied by Özdönmez et al. (2018). The distribution peaks at $t_3 \simeq 30$ d, which is roughly the line of demarcation between the so-called "fast" and "slow" novae.

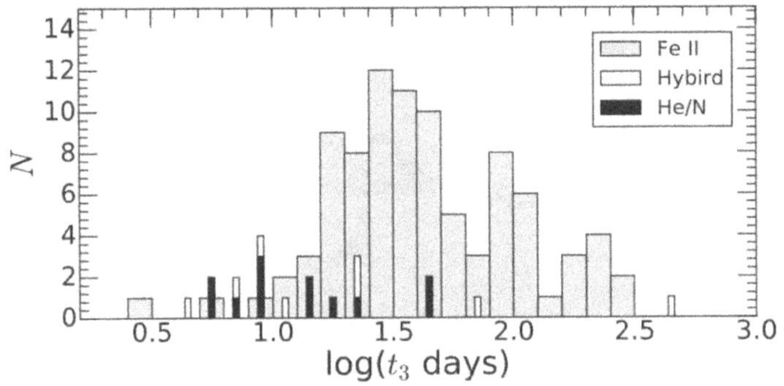

Figure 8.4. The distribution of the t_3 times for Galactic novae with known spectroscopic class. (Reproduced from Özdönmez et al. 2018. Copyright of OUP © 2018.)

8.2 The MMRD Relation Revisited

No other aspect of nova behavior has been more heavily scrutinized in recent years than has the maximum-magnitude, rate-of-decline relation discussed in Chapter 2. After its introduction in 1945 (McLaughlin 1945), the MMRD relation was widely accepted, and often used for determining distances to novae where peak apparent magnitude and fade rate could be measured. In recent years, the recognition that recurrent novae, whose outbursts are typically less luminous and faster developing than other novae, do not fit the traditional MMRD relation has caused the entire relationship to be questioned. Before discussing the latest observations, we pause briefly to summarize the historical development of the MMRD.

8.2.1 Historical Development

The MMRD is an empirical relationship between a nova's peak absolute magnitude and its rate of decline. The relation is usually cast in the form: $M_V = a + b \log t_n(V)$, where a and b are the best fitting intercept and slope of a linear fit to the data, and t_n is the time in days it takes a nova to fade n magnitudes from peak (n is typically taken to be 2 or 3). Unlike the behavior of Type Ia supernovae, where the more luminous systems fade the slowest, luminous novae have traditionally been thought to fade more rapidly on average than their less luminous counterparts.

The MMRD relation has been re-calibrated and re-assessed many times over the years. In 1957, Theodor Schmidt (1957) recalibrated the relation and used it to determine a distance modulus, $\mu_0 = 24.4$ for the Andromeda galaxy—remarkably consistent with the currently accepted value of $\mu_0 = 24.38 \pm 0.05$ (Freedman et al. 2001). The status of the MMRD some 15 years after its introduction was reviewed by McLaughlin (1960) who found that the relation $M_V = -11.5 + 2.5 \log t_3(V)$ provided the best fit to the available Galactic data.

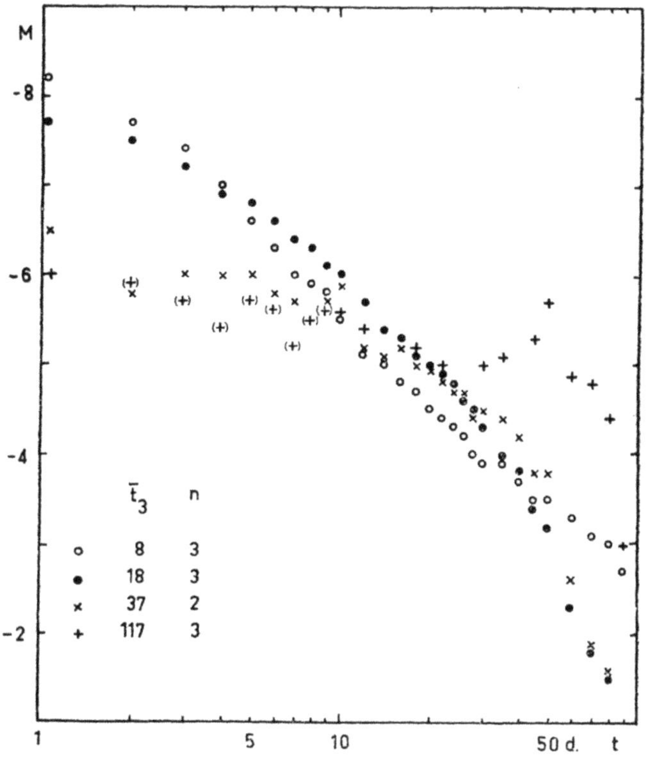

Figure 8.5. The MMRD relations for four groups of novae with mean fade rates characterized by $\langle t_3 \rangle = 8$, $\langle t_3 \rangle = 18$, $\langle t_3 \rangle = 38$, and $\langle t_3 \rangle = 117$ d. The dispersion is a minimum at 15 days post maximum where $\langle M \rangle = -5.2 \pm 0.1$. (Reproduced from Buscombe & de Vaucouleurs 1955.)

The Buscombe–de Vaucouleurs Relation

In 1955 Buscombe & de Vaucouleurs (1955) calibrated the MMRD relation using 11 well observed Galactic novae and found a best-fitting relation of $M_V = -10 + 2.3 \log t_3$. Noting that since brighter novae faded more quickly than less luminous novae, they reasoned that there should be a time after eruption when all novae reach a similar luminosity. An inspection of the data revealed that the dispersion in nova luminosities was a minimum approximately 15 days after peak luminosity, where novae had a characteristic luminosity of $M_{15} = -5.2 \pm 0.1$ (see Figure 8.5). Through a comparison with light curve data of 10 novae in the Magellanic Clouds, using their "M_{15}" relation, Buscombe and de Vaucouleurs derived a distance modulus to the Clouds of $m - M = 18.6$.

Like the MMRD, the Buscombe–de Vaucouleurs relation has been recalibrated several times over the years, and it is now understood that the original calibration overestimated the value of M_{15} by approximately 1 mag. Most recently, Shara et al. (2018) have analyzed the light curves of 41 M87 novae obtained with the *Hubble Space Telescope* and suggested that $M_{15} = -6.37 \pm 0.46$ and $M_{15} = -6.11 \pm 0.43$ for the F606W and F814W filters, respectively. However, as we will explore below, a major caveat is that the relation does not apply to all novae! Specifically, a class of co-called "faint and fast novae," (FFN) do not obey this relation.

Over the next several decades the MMRD relation was re-calibrated several more times, most notably by Judith Cohen (1985) and later by Downes & Duerbeck (2000). Both of these authors took care to only include novae with the most accurate distances (and hence absolute magnitudes) in their samples. The MMRD relation found by Cohen (1985) is shown in Figure 8.6. The left panel shows the classical MMRD that results from adopting the values of M_V(max) for each nova individually. The modified MMRD shown in the right panel was constructed by shifting the absolute magnitudes at maximum light to force all light curves to obey a Buscombe–de Vaucouleurs relation—in this case to reach a constant absolute magnitude of $M_V = -5.60 \pm 0.14$ (the mean of the individual novae) 15 days after peak brightness. Clearly, the MMRD fit is improved when a Buscombe–de Vaucouleurs relation is enforced, but it is unclear if this can be justified *a priori*.

8.2.2 Sources of Uncertainty

Acquiring reliable data to formulate an MMRD relation is often challenging. Two quantities must be measured—the absolute magnitude at maximum light, and the time it takes the nova to fade by a specific number of magnitudes (typically 2 or 3 mag)—neither of which is straightforward. Let's start by considering the absolute magnitude at maximum light. The brighter Galactic novae are often discovered prior to maximum light, making a determination of the time and apparent magnitude at maximum light possible. For novae with resolved shells, distances have been traditionally measured through expansion parallax. Expansion parallax

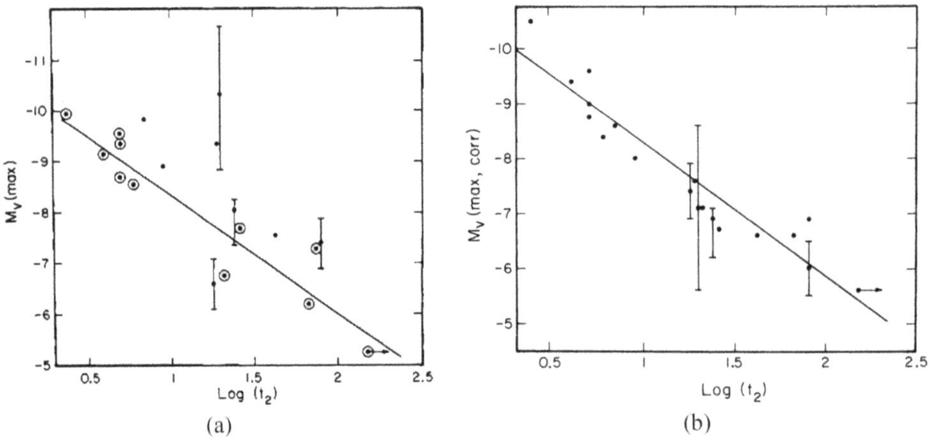

Figure 8.6. (a) The MMRD relation from a sample of Galactic novae studied by Cohen (1985). Circled points represent novae with distances (and thus absolute magnitudes) that have been determined through expansion parallax of the nova shell. (b) The MMRD relation from (a) with individual data points shifted so that all novae have the same absolute magnitude (in this case $M_V = 5.60$) 15 days after peak brightness (i.e., they obey the Buscombe–de Vaucouleurs relation; Buscombe and de Vaucouleurs 1955). (Reproduced from Cohen 1985.

has the advantage that it is a geometrical technique for distance determination that in unaffected by extinction, however a number of assumptions go into its application. For example, the nova shells are usually assumed to be spherical and expanding at a constant rate since the eruption. Nova shells can have complex morphologies, and are generally prolate rather than spherical. Depending on the orientation to our line-of-sight the assumption of spherical symmetry will introduce errors in the distance determinations.

Just recently, *Gaia*[1] parallaxes have improved the distance estimates for a large sample of Galactic novae. However, to compute the absolute magnitude at maximum light requires that the extinction be determined. Since most novae are concentrated to the Galactic plane, extinction is often not negligible, and this has been a persistent problem in deriving accurate absolute magnitudes. As a result the scatter in the MMRD relation has been large, and it has been difficult to disentangle the intrinsic scatter in the relation from the observational uncertainties. To make matters worse, estimating the time to decline by 2 or 3 mag from maximum light (i.e., t_2 or t_3) is not a trivial exercise. The difficulty can be appreciated by looking at the diversity of light curves shown in Figure 8.3. Nova lightcurves are extremely inhomogeneous, with flares, oscillations and dips that make consistent measurements of t_2 or particularly t_3 times virtually impossible in all except novae that fall into the "smooth" (S) category.

Unfortunately, there doesn't appear to be much one can do about making measurements of nova decline rates consistent across the range of light curve morphologies. Perhaps, given that most light curves seem to decline smoothly within the first ~2 mag from maximum light, using t_2 would seem preferable to adopting the longer t_3 times. Another option might be to restrict MMRD samples to novae belonging to the smooth "S" class. On the other hand, reducing the scatter in the absolute magnitudes at maximum light should be possible by observing a large sample of roughly equidistant novae in a nearby galaxy. Although the Magellanic Clouds served their role well in the case of Cepheid variables (Leavitt & Pickering 1912), the nova rates in the Clouds are too low to be practical in this regard. The Andromeda galaxy, however, is almost ideal, the high inclination of the Galaxy to our line of sight notwithstanding.

As part of their spectroscopic survey of novae in M31, Shafter et al. (2011a) included light curves for a sub-sample of 47 novae with available photometric data. Fade rates parameterized as t_2 times were measured by least-squares fitting to the declining portion of the light curve, with no attempt made to distinguish between the different possible Strope light-curve classes. The data were presented graphically in the form of MMRD plots for B, V, and R band measurements. The *R*-band data were the most extensive, and the corresponding MMRD plot is shown in Figure 8.7. While a general tendency exists for fast fading novae to be more luminous than their slower declining counterparts, the relationship is hardly impressive.

[1] *Gaia* is a European Space Agency space observatory designed to measure parallaxes of more than a billion stars down to $m \sim 20$. It was launched in 2013 and is expected to operate until approximately 2022 or 2024.

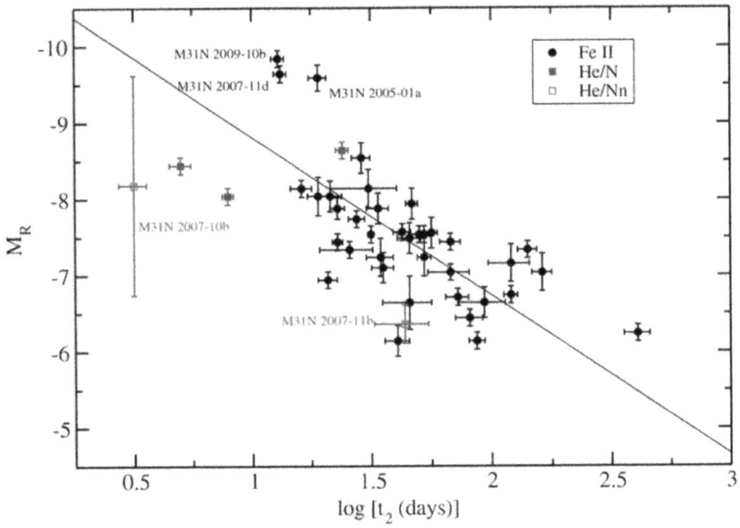

Figure 8.7. The MMRD relation for a sample of M31 novae with well sampled light curves. (Reproduced from Shafter et al. 2011a. © 2011. The American Astronomical Society. All rights reserved.)

8.2.3 The Skepticism Mounts

Its long history notwithstanding, confidence in the MMRD relation as a useful tool for measuring extragalactic distances has never been particularly high. A major concern has been that despite the acquisition of more precise data over the years, and the continued attempts at recalibration, the scatter in the relation has not diminished significantly. Theoretical models have shown that there are multiple parameters governing the outburst characteristics of novae, and that the maximum luminosity and fade rate are not simply related. Thus, it became gradually accepted that the MMRD relation would never be a precision distance indicator on a par with, for example, the PLC relation for Cepheid variables. In recent years, the tide has turned sharply against the MMRD, with the opposition approaching an almost religious fervor. Not only is the MMRD being rejected as a potentially useful distance indicator, but the very existence of the relation is now being questioned.

This dramatic turnaround was triggered by the discovery of several apparently faint but fast-declining novae in M31. During the run up to the formal start of PTF operations, a program P60-FasTING (Palomar 60 in, Fast Transients in Nearby Galaxies) was undertaken by Mansi Kasliwal as part of her doctoral dissertation at Caltech. The survey had a fast (1 day) cadence and it included several nearby galaxies to a depth of $g \leqslant 21$. In an important paper, Kasliwal et al. (2011) published a study from the P60-FasTING program in which they studied the peak magnitudes and luminosities of a sample of M31 novae, finding evidence for a population of FFN that appeared to violate the iconic MMRD relation (see Figure 8.8).

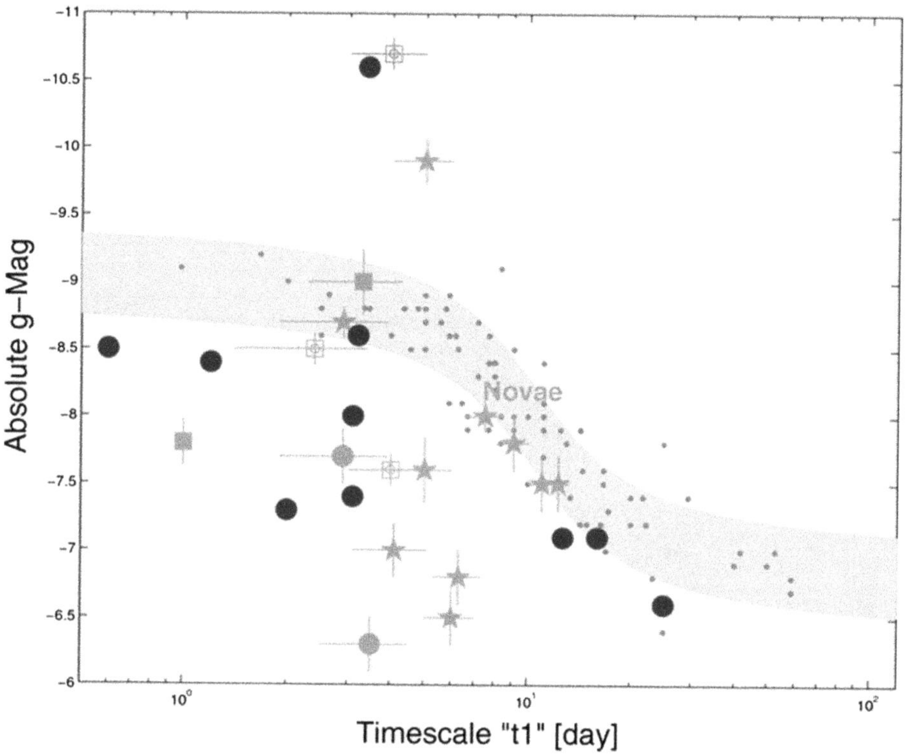

Figure 8.8. An MMRD relation comparing novae found in the P60-FasTING survey (green star symbols) with Galactic recurrent novae (blue filled circles). Clearly, many of the novae fail to follow the classic MMRD relation from Valle & Livio (1995) indicated by the shaded region. (Reproduced from Kasliwal et al. 2011. © 2011. The American Astronomical Society. All rights reserved.)

8.2.4 The Latest Results for the MMRD

Two studies published in 2018, and a third in 2019, have made significant strides in characterizing the Galactic MMRD relation. The first of these was conducted by Özdönmez et al. (2018), who used distances and absolute magnitudes from the large sample of Galactic novae the group had compiled two years earlier (Özdönmez et al. 2016). Of these, a subsample of 65 novae with the most reliable distances were used to produce the MMRD relations shown in Figure 8.9. The data are separated into two groups with distances either estimated from expansion parallax, or through the use of reddening–distance relations (RDRs) made possible through IR surveys of red clump giant stars (of known luminosity) having lines of sight near that of a target nova. As can be seen from an inspection of Figure 8.9, nova luminosity appears to be weakly correlated with the rate of decline from maximum light (inversely correlated with log t_3). The MMRD correlation is strongest when only novae with expansion parallax distances are included, but even then, there is considerable scatter in the relation.

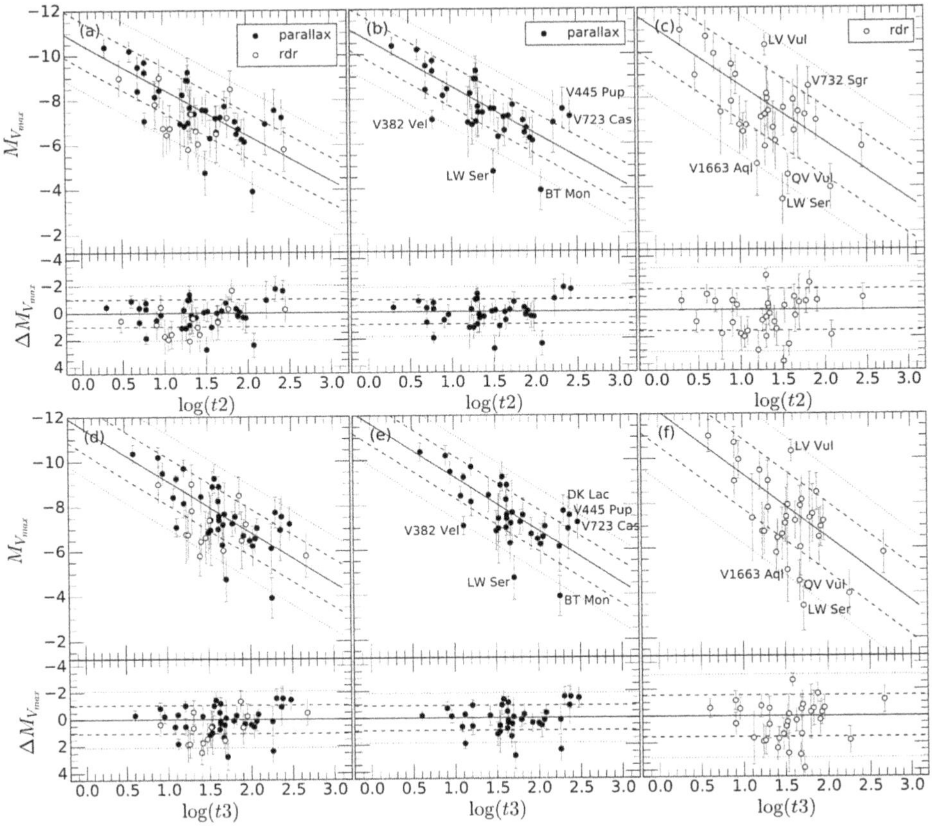

Figure 8.9. Galactic MMRD relations based on t_2 (top) and t_3 (bottom) from the study of Özdönmez et al. (2018). The left, middle and right columns show novae with distances based on expansion parallax plus the RDR, expansion parallax alone, and the RDR alone, respectively. The solid lines are the best fit MMRD relations, with the dashed and dotted lines representing 1σ and 2σ deviations, respectively. (Reproduced from Özdönmez et al. 2018. Copyright of OUP © 2018.)

Shortly thereafter a thought-provoking article was published by Bradley Schaefer, who explored an MMRD relation based on distances derived using data from the *Gaia* mission (Schaefer 2018). Parallaxes for a total of 64 novae were included in the second *Gaia* data release (DR2). For purposes of evaluating prior methods for obtaining nova distances, and ultimately for evaluating the utility of the MMRD, Schaefer divided the 64 novae into three groups: (1) A "gold" group consisting of novae with parallax errors <30% *and* well-determined light curves suitable for measuring decline rates, (2) a "silver" group with parallax errors also <30%, but with less favorable light curve information, and (3) a "bronze" group with parallax measurements having uncertainties >30%. A total of 41 novae had sufficient quality data to be included in the gold and silver samples.

Based on a comparison of the *Gaia*-based distances with those from expansion parallaxes and the RDR method of Özdönmez et al. (2016, 2018). Schaefer concludes that only distances based on *Gaia* parallaxes and the RDR method are

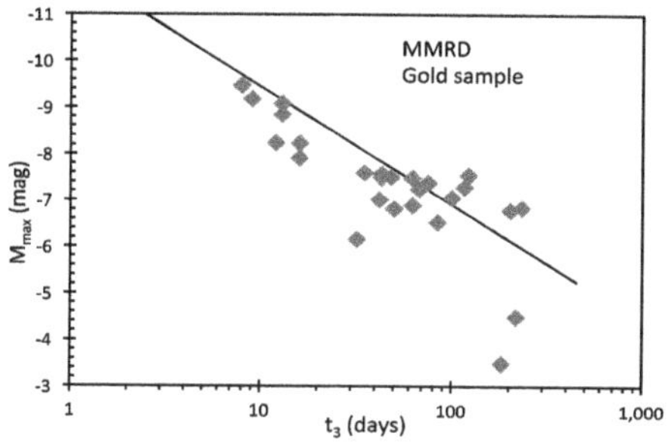

(a)

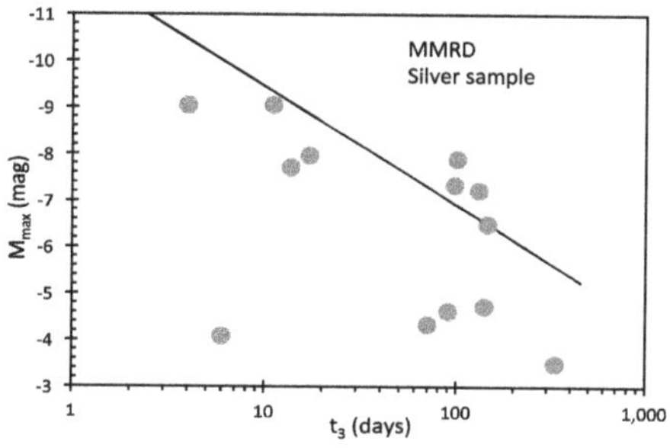

(b)

Figure 8.10. MMRD relations based on *Gaia* parallaxes. The left panel (a) shows the "gold" sample data with the highest quality photometric parameters, while the right panel shows the MMRD based on lesser quality "silver" sample photometric data. (Reproduced from Schaefer 2018. Copyright of OUP © 2018.)

sufficiently accurate to warrant application to individual novae. Surprisingly, the one method thought to be the "gold standard" for Galactic nova distances, expansion parallax of the nova shell, was found by Schaefer to be particularly poor. The comparison of expansion parallax distances with *Gaia* distances results in a correlation with, in Schaefer's words "a horrifyingly large scatter."

Figure 8.10 shows the resulting MMRD relations for the gold and silver samples. Although a clear correlation exists for the gold sample, Schaefer argues persuasively that the scatter in the MMRD relations is sufficiently large that they are unsuitable for nova distance determinations. Specifically, he concludes: "The real scatter is so large that the MMRD (if it exists at all) is useless for studies of individual novae and even useless for statistical purposes. From this, I conclude that the MMRD has completely failed as a prior method to get nova distances." When parallax measurements or RDR measurements are not available, and a distance is required, he recommends simply taking the mean nova absolute magnitude at maximum light $M_V = -7.0 \pm 1.4$ as a last resort.

The most recent study of the MMRD, also based on *Gaia* parallaxes, has been reported by Selvelli & Gilmozzi (2019). These authors considered a total of 18 novae from an earlier study with UV spectra sufficient to derive accretions rates during quiescence (Selvelli & Gilmozzi 2013), that now have well-determined *Gaia* distances. In contrast to the conclusions reached by Schaefer a few months earlier, Selvelli and Gilmozzi found a rather tight correlation between the luminosity at maximum light and the subsequent rate of decline. Their MMRD relation is shown in Figure 8.11.

In summary, MMRD relations based on the latest *Gaia* distances have yielded conflicting results, at least in terms of interpretation. A definitive answer may have to await subsequent *Gaia* data releases where the uncertainties due to the effects of binary motion, for example, should be significantly reduced (Lindegren et al. 2018). In the mean time, it remains unclear how much of the observed scatter in the MMRD relation is intrinsic to the novae themselves and how much is a result of uncertainty in parameters such as extinction, both in the nova shells and along the

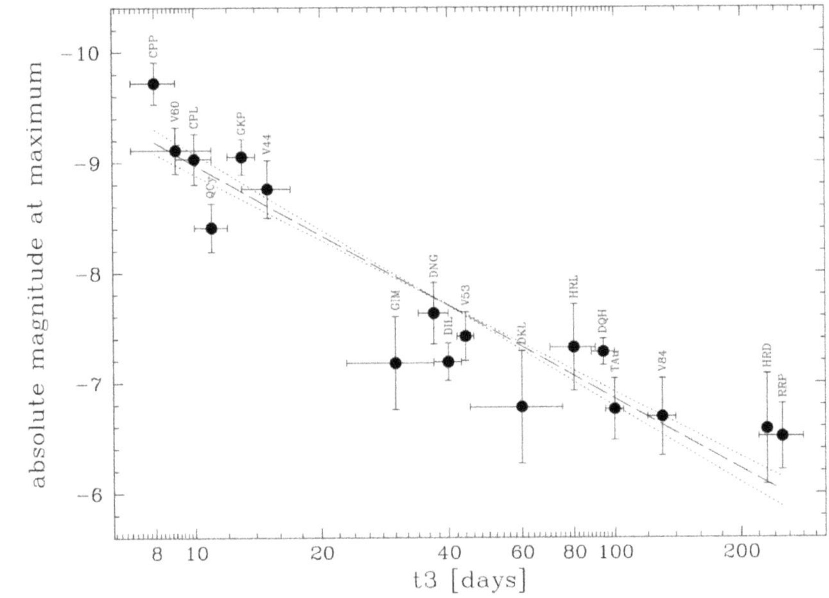

Figure 8.11. The MMRD relation based on 18 novae from the recent study by Selvelli and Gilmozzi. The peak nova luminosities from this GAIA sample are strongly correlated with their rate of decline. (Reproduced with permission from Selvelli & Gilmozzi 2019, © ESO.)

line of site to the individual novae. The general consensus seems to be that the intrinsic scatter is sufficiently large that the MMRD is not particularly useful for obtaining accurate distances to novae. That said, the available evidence certainly supports the notion that a rough correlation between a nova's luminosity at maximum light and its subsequent rate of decline does indeed exist, with the more luminous novae fading faster on average compared with their less luminous counterparts. In this limited sense, the MMRD still survives.

8.3 The Nature of the Faint and Fast Nova Population

Perhaps the most important question regarding nova physics that remains unresolved concerns the size of the population of FFN. Clearly, the existence of a significant FFN population is incompatible with a universal MMRD relation for novae. Furthermore, it must be incompatible with a Buscombe–de Vaucouleurs relation as well, since the latter is predicated on the existence of a MMRD relation. In order for novae to reach a common absolute magnitude at a fixed time post eruption, the brighter novae must fade faster than their less luminous counterparts. If novae don't obey the MMRD relation, they won't obey a Buscombe–de Vaucouleurs relation. Regardless of how significant the FFN nova population turns out to be, there is no doubt that such systems have been strongly selected against in past nova surveys. Even if FFN turn out to be quite common, it is easy to understand how the MMRD survived for so long. It will have resulted in being simply a selection effect favoring the discovery of brighter and slower evolving novae.

The notion that FFN had been simply overlooked in earlier surveys seemed quite compelling given that most modern M31 nova surveys relied on low-cadence observations given the pressures on telescope time. In addition, since many recent surveys were based on Hα imaging (e.g., Ciardullo et al. 1987; Shafter & Irby 2001; and the RBSE study). Because novae fade more slowly in Hα, it was not believed necessary to image as frequently to discover most novae. The case for the putative FFN population had its genesis with the discovery in 2008 of a most remarkable recurrent nova in M31.

8.3.1 The Remarkable Recurrent Nova M31N 2008-12a

A few years before Kasliwal's P60-FasTING survey would surprise the nova community with the discovery of a sample of faint and fast novae in M31, a couple of dedicated and energetic amateur astronomers from Japan, Koichi Nishiyama and Fujio Kabashima, were in the process of conducting their own M31 nova patrol using a 50 cm telescope at their Miyaki-Argenteus Observatory (see Figure 8.12). Well known within the professional ranks, Nishiyama and Kabashima have discovered a total of more than 100 novae in M31 over the past decade. Perhaps their most noteworthy discovery, albeit not appreciated at the time, was made on the night of 2008 December 26, when they spotted a faint ($m \sim 18.5$) transient in the outskirts of M31, approximately a full degree North–North–East of the nucleus. The object, M31N 2008-12a, which faded quickly and was not confirmed spectroscopically, soon fell into relative obscurity.

Then, unexpectedly, almost three years later a pair of Russian amateur astronomers Stanislav Korotkiy and Leonid Elenin discovered a faint transient ($m \sim 18.6$) at the position of 2008-12a on the night of 2011 October 22 using the 0.45 m astrograph at the ISON-NM Observatory, Mayall, New Mexico (Barsukova et al. 2011). The elusive object was quickly confirmed on the following night by Kamil

(a)

(b)

Figure 8.12. (a) The Miyaki-Argenteus Observatory in Miyaki, Saga Prefecture, Japan. (b) The 50 cm reflector used by Koichi Nishiyama and Fujio Kabashima at the Miyaki-Argenteus Observatory. (Images courtesy of Fujio Kabashima.)

Hornoch with observations from the Ondřejov observatory. Regarding its authenticity, in an email to the author, Hornoch was unequivocal: "… M31N 2011-10e is a real object, ⋯ it probably belongs to group of fast and faint novae, because it was not probably brighter than about 18.5 mag and faded down to 20 mag or so in two-three days …." Additional photometric observations were obtained approximately three days after eruption by Barsukova et al. using the large 6 m BTA telescope in the Caucasus mountains. Data acquired on October 26.97 showed that the object had faded to $V = 21.2 \pm 0.16$ and $B = 20.9 \pm 0.15$.

The prospect that M31N 2008-12a could be a recurrent nova with a recurrence time of just three years sparked considerable interest. Despite Hornoch's prescient speculation, it was unclear at the time if the amount of material required to trigger a TNR could be accreted in such a short time, given that the mass accretion rate cannot be arbitrarily high, but must remain sufficiently low during the interruption interval to maintain at least partial degeneracy in the accreted envelope. Models of TNRs on accreting white dwarf stars had not carefully explored this extreme high $M_{\rm WD}$, high $\dot{M}$ region of parameter space. The shortest recurrence time known for any nova was for M31N 1963-09c, a nova discovered as part of the Rosino et al. (1973) survey which had been seen to erupt five years later in September of 1968. In the Galaxy, only U Sco had a recurrence time of less than 10 years.

Perhaps the observed eruptions of M31N 2008-12a and 2011-10e weren't independent eruptions, perhaps they were part of the same slow nova eruption with the 2011 event being a re-brightening of the 2008 eruption, or perhaps the object wasn't a nova at all, but rather a long period variable star. It wouldn't have been the first time that a long-period variable star had been caught masquerading as a nova in M31 (e.g., see Shafter et al. 2008). In the case of 2011-10e, though, that possibility seems unlikely given R-band plus narrow-band Hα observations by Hornoch which indicated that the object had Hα in emission. Nevertheless, the only way to resolve the issue was to obtain a spectrum of the object. Unfortunately, M31N 2008-12a is a faint target, even at maximum light, and it fades rapidly. Shafter and collaborators attempted to acquire a spectrum with the McDonald Observatory's HET on the night of 2011 October 29, a week after the discovery of M31N 2011-10e. Unfortunately, by that time, 2011-10e has faded to the point where identification of the target became extremely difficult (Shafter et al. 2011b; Shafter 2011).

The next opportunity to obtain a spectrum of 2008-12a would come surprisingly soon, a little less than a year later when M31N 2008-12a erupted yet again on 2012 October 18. The eruption, designated M31N 2012-10a, was discovered by Nishiyama and Kabashima, the same amateur astronomers who had discovered the first eruption back in 2008 December. This time, spectroscopic observations would be forthcoming. Less than two days post discovery, a low-dispersion, optical spectrum was obtained with the HET by Shafter et al. (2012). The spectrum, shown in Figure 8.13, displaying strong and broad emission lines of H, He, and N, was clearly that of a nova, although somewhat atypical.

Normally, novae displaying significant He and N emission shortly after eruption are characterized by extremely broad Balmer emission, of order 5000 km s^{-1}.

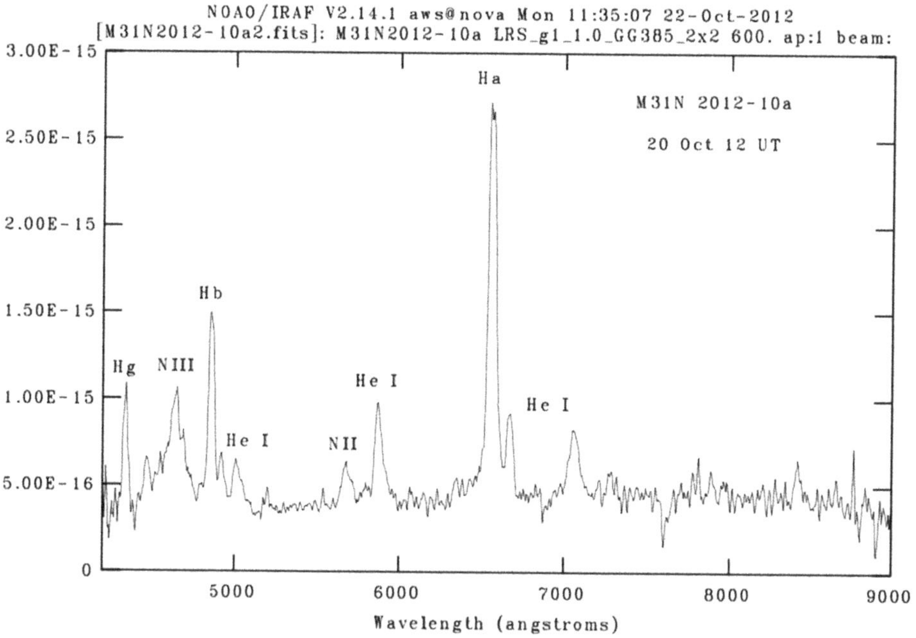

Figure 8.13. An early spectrum of the remarkable recurrent nova M31N 2008-12a obtained with the HET. Note the broad emission lines of H, He, and N, typical of an He/N class nova. Ordinate is F-lambda.

Such systems are members of the He/N spectroscopic class as described earlier. Most recurrent novae are members of this spectroscopic class. In the case of 2012-10a, the lines were relatively narrow, of order 2500 km s^{-1}, which is at the upper end of the velocities typically observed in the other main class of novae, the Fe II systems. If 2012-10a was indeed a recurrence of M31N 2008-12a, its ejection velocity was lower than those of typical He/N novae. In addition, if 2012-10a was a short recurrence time RN, then a brief supersoft phase should ensue with a few week after eruption. To test this prediction, a Swift proposal was put together by Michael Bode, with observations initiated on 2012 November 5, with follow-up observations two weeks later, on November 16, and then again on December 16. The system was detected (marginally) in the first observation, but not on the subsequent observations.

At the close of 2012, the true nature of M31N 2012-10a was still unclear. Spectroscopy had ruled out a long-period variable star, but was it possibly a foreground cataclysmic variable of some sort, a slow nova in M31 that had several re-brightenings, or a recurrent nova in M31 with the shortest period between eruptions ever observed, (by a factor of at least 5!)? The situation was summarized by Shafter et al. (2012) in their Astronomer's Telegram from 2012 October: "… the spatial coincidence with M31N 2008-12a and 2011-10e suggests that the object is a recurrent nova, but given the apparently low luminosity and the unusually short interval between brightenings, it is unclear whether the current outburst represents a new eruption, or a re-brightening of the recent 2011 event". Although existing models for thermonuclear runaways on accreting white dwarfs had not explored in

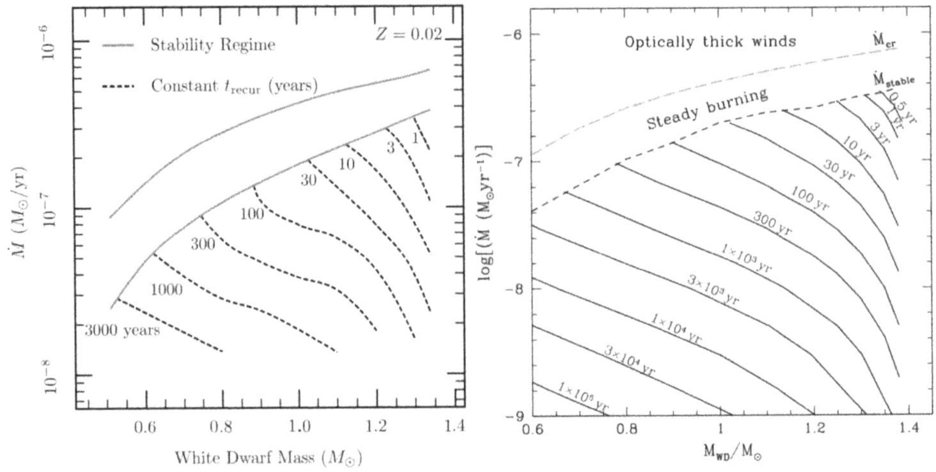

(a) (b)

Figure 8.14. "Nomoto diagrams" showing how the recurrence times of novae depend on the white dwarf's mass and accretion rate. To achieve a recurrence time of order one year, as in the case of M31N 2008-12a, white dwarfs with masses near the Chandrasekhar limit accreting above 10^{-7} $M_\odot$ yr^{-1} are required. (Reproduced from Wolf et al. 2013 and Kato et al. 2014. © 2013 and 2014, respectively. The American Astronomical Society. All rights reserved.)

detail the parameter space associated with extremely short recurrence times, it was clear that if 2008-12a was a RN, the accretion rate must be very high, near the minimum value for steady burning, and the white dwarf mass must be approaching the Chandrasekhar limit.

Prompted by the extremely rapid recurrence time exhibited by 2008-12a, (Wolf et al. 2013) at the University of California, Santa Barbara used the Modules for Experiments in Stellar Astrophysics (MESA) code (Paxton et al. 2011, 2013) to examine the properties of white dwarfs accreting at high rates near the stable burning regime. They were specifically interested in exploring the relationship between the recurrence time and white dwarf mass in recurrent nova systems. Figure 8.14 shows the relationship between mass accretion rate, white dwarf mass, and recurrence time from models by Wolf et al. (2013) and Kato et al. (2014).

Given the back-to-back eruptions (or re-brightenings?) observed in 2011 and 2012, the nova community was fully prepared to observe the eruption expected to occur in the fall of 2013. On the night of 2013 November 27 UT the next eruption of M31N2008-12a was discovered by the iPTF collaboration (Tang et al. 2013). Shortly thereafter, Liverpool John Moores University doctoral student Steven Williams and collaborators reported an *HST* archival study of the 2008-12a field, and the discovery of the progenitor system, a star with $I = 23.90 \pm 0.03$ and $B - I = 0.22 \pm 0.03$ lying 8 mas (0.9σ) from the position of M31N 2013-11-f (Williams et al. 2013). If this represented the true quiescent state, then M31N 2008-12a was indeed a ~1 yr recurrent nova in M31 with an eruption amplitude of ~6 mag. As pointed out by Williams et al. at the distance of M31, the quiescent system would have an absolute magnitude $M_B \simeq -0.05$, which is similar to the Galactic recurrent novae containing red giant secondaries such as V2487 Oph, and subgiant secondaries such as U Sco. The relatively blue color being due to the assumed high mass accretion rate onto the white dwarfs in these systems.

Despite the fact that the pieces were falling into place behind the notion that M31N 2008-12a was indeed a recurrent nova in M31, with an unprecedented one-year recurrence time, the (remote) possibility that M31N 2008-12a was a slow nova with multiple "flare-ups" on a single extended eruption was still considered viable, and caution was exercised when describing the system. For example, the title of the iPTF collaboration ATel read "A Candidate Recurrent Nova in M31 from iPTF," and Williams et al. ended their ATel with the caveat that "Given the apparent short outburst interval of this nova it is possible that we are simply observing part of the late decline following the 2008 outburst."

Everything changed in the opening days of 2014 with the contemporaneous publications by two independent groups of two peer-reviewed publications announcing that 2008-12a was indeed a remarkable one year recurrence time nova in M31. First to appear on the arXiv preprint server was a manuscript entitled "An Accreting White Dwarf near the Chandrasekhar Limit in the Andromeda Galaxy" by a Caltech—University of California Santa Barbara collaboration led by Tang et al. (2014). That work was followed shortly thereafter by a pair of papers by Matthew Darnley, Martin Henze, and their collaborators entitled "A remarkable recurrent nova in M31: The optical observations" (Darnley et al. 2014) and "A remarkable recurrent nova in M31: The X-ray observations" (Henze et al. 2014).

The fact that M31N 2008-12a must harbor a near-Chandrasekhar mass white dwarf accreting at a rate above $10^{-7}\ M_{\odot}\ yr^{-1}$ raises the interesting possibility that the system is a prime candidate for a Type-Ia supernova progenitor (Tang et al. 2014; Kato et al. 2014). Tang et al. using the MESA code, have shown that the white dwarf retains of order 30% of the mass it accretes, and they predict that the white dwarf will reach the Chandrasekhar limit in a time of order 5×10^5 years. What happens next depends on the composition of the white dwarf. A ONe core would be expected to undergo an accretion-induces collapse (AIC) to a neutron star, while a CO core is expected to undergo a deflagration resulting in a SN Ia event.

Given the possibility that M31N 2008-12a may be the most promising SN Ia progenitor extant, there has been a sustained interest in this nova in recent years, mostly by an international team led by Matt Darnley at Liverpool John Moores University and Martin Henze, first at *ICE* Barcelona, and then at San Diego State University. As of the end of 2018, a total of 11 eruptions have now been observed—one each year since 2008 December. Discoveries include a fossil shell surrounding the nova formed by multitudes of prior eruptions (see Figure 8.15).

Despite the P60-FasTING study and the contemporaneous discovery of M31N 2008-12a, the question of whether a significant FFN population exists is still unclear. It's possible that some of the FFN discovered in the P60-FasTING survey were in fact more luminous novae that were heavily extincted, although this would seem unlikely to apply to all of these novae. Subsequent analysis of PTF and other high-cadence surveys has been limited. As mentioned previously, Soraisam and

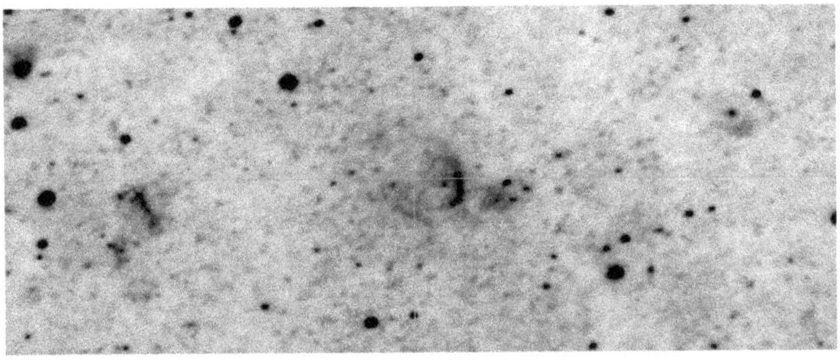

Figure 8.15. An Hα image of the field of M31N 2008-12a obtained on 2005 October 8 by A. W. Shafter and K. A. Misselt with the Steward Observatory 90 inch reflector. The image re-analyzed years later by Shafter and former graduate student David Baer led to the discovery in 2014 March that the position of the nova (marked by the cross hairs) appeared to be surrounded by an nebular shell, seen most prominently to the right and below the position of the nova.

collaborators (Soraisam et al. 2017) studied a limited sample of data from the PTF survey and found only two novae that might be examples of FFN. Clearly, additional work must be undertaken in the analysis of nova light curves from deep, high-cadence surveys before we can assess the significance of a putative FFN nova population.

8.4 Recurrent Nova Populations

Although all novae are recurrent on some timescale, systems with recurrence times less that of order a century that have been observed to erupt more than once are collectively known as recurrent novae (RNe). There are a total of 10 RNe known in the Milky Way, and their properties have been thoroughly reviewed by Bradley Schaefer in his now classic paper published in 2010 (Schaefer 2010). These 10 RNe account for 37 of the roughly 450 recorded nova eruptions in the Milky Way. Thus, ~8% of the *observed* nova eruptions in the Galaxy are from known RN systems. Of course, this is a strict lower limit on the fraction of all outbursts that arise from RNe, as recurrent novae must be discovered more than once. Thus, the discovery probability for RNe scales roughly as the square of the discovery probability for a typical nova.[2] Bradley Schaefer and Ashley Pagnotta considered several aspects of Galactic nova discovery probabilities and estimated that roughly 25% of the observed nova outbursts in the Milky Way actually arise from RN systems (Pagnotta & Schaefer 2014).

Outside the Galaxy, a total of 22 recurrent novae have been identified, 18 of which have been found in the nearby Andromeda galaxy (Shafter et al. 2015; Hornoch & Shafter 2015; Sin et al. 2017; Socia et al. 2018). The four remaining RN systems are all situated within the LMC (Shafter 2013; Bode et al. 2016; Mróz & Udalski 2018). Among well studied galaxies, no RN candidates have yet been

[2] Actually, it is a bit more complicated than this because recurrent novae with very short intervals between eruptions provide multiple opportunities to discover the recurrent nature of the system.

identified in the SMC, M33, and perhaps surprisingly in M81, where more than 220 nova eruptions have been observed.

The fraction of novae in the LMC that are recurrent appears to be remarkably high (Shafter 2013). Under the assumption that both the LMC and M31 have been monitored with approximately the same frequency, RNe appear to be about an order of magnitude more common in the LMC compared with M31 after normalization by the galaxys' K-band luminosities. In terms of the number of outbursts, approximately 4% of the eruptions observed in M31 have come from identified recurrent nova systems (Shafter et al. 2015), whereas outbursts associated with known recurrent novae account for 25% of all observed LMC nova eruptions. Thus, outbursts from known recurrent nova progenitors are ~6 times more frequent in the LMC compared with M31. As with the Galactic RNe, the raw observed RN rates must be corrected for discovery efficiencies. In their 2015 paper on RNe in M31, Shafter et al. used numerical simulations to estimate that as many as one in three nova eruptions in M31 may arise from RN systems. An analogous determination of the discovery efficiency of the LMC observations have not been done, but judging by the corrections found for the Galaxy and M31, it seems likely that the majority of the nova eruptions observed in the LMC come from RNe.

In their OGLE study, Mróz et al. (2016) noted the relatively high $LSNR_K$ in the LMC and attributed it to the Galaxy's star formation history. In particular, they argue that the bulk of the LMC novae were the product of a second burst of star formation approximately 3–5 Gyr ago. As shown by Yungelson et al. (1997), such a young population of novae should contain generally more massive white dwarfs compared with novae arising from older stellar populations. The more massive white dwarfs in turn will result in more frequent nova eruptions from a given progenitor binary, more RNe, and a higher overall $LSNR_K$.

8.4.1 Recurrent Novae as Progenitors of Type Ia Supernovae

Despite being binary systems containing an accreting white dwarf, novae have traditionally been considered poor candidates for the progenitors of Type Ia supernovae. The main objection has been the belief that mass loss in nova eruptions will prevent the white dwarfs from growing in mass and eventually reaching the Chandrasekhar limit. However, recent work by several groups have shown that under the right conditions the white dwarf mass can continue to grow despite repeated nova eruptions (Hillman et al. 2016; Kato et al. 2014; Starrfield et al. 2012). The one year recurrence time nova M31N 2008-12a is a good example of a system where the white dwarf mass is increasing (Tang et al. 2014).

An obvious question to ask is what fraction of Type Ia supernovae (SNe Ia) could in principle be produced by RNe. This question was explored recently by Soraisam & Gilfanov (2015) who used multi-cycle nova evolutionary models from Yaron et al. (2005) to predict the number and frequency of nova eruptions expected from a typical SNe Ia progenitor system. A comparison with the observed nova rate in M31 of 65 yr^{-1} led Soraisam and Gilfanov to conclude that only 2%–7% of the estimated SNe Ia rate in M31 could be produced by relatively slow novae with $t_2 > 10$ d characteristic of the POINT-AGAPE survey. Shafter et al. (2015) arrived at a similar estimate from an analysis of the completeness of the observed M31 RN population.

As a population of novae evolves and their white dwarfs approach the Chandrasekhar limit, the resulting outbursts will become more frequent, and the t_2 times shorter. Thus, if novae are to contribute significantly to the SNe Ia rate, we would expect FFN to dominate a galaxy's nova population. As we have seen, there is little evidence for this putative population in extant nova surveys. Clearly, we must definitively characterize the FFN population through future deep, high-cadence surveys before we can fully evaluate the role that novae may play as a significant channel for the production of SNe Ia.

8.5 Outstanding Questions Concerning Nova Populations

As explored in the previous chapter, the question of whether the observed properties and rates of novae vary with stellar population across the Hubble sequence remains uncertain. Clearly, a major problem has been the large uncertainties in the derived luminosity-specific nova rates. Before any further progress can be made on this question, and on our understanding of the MMRD relation for that matter, a more reliable determination of the nova luminosity function, which is basic to our understanding of many key aspects of nova physics, must be realized.

There were early indications that the luminosity-specific nova rates in later Hubble type galaxies (M33 and the Magellanic Clouds) were higher than the rates in earlier type galaxies (e.g., see, Della Valle et al. 1994). However, recent observations have shown that the LSNRs in elliptical galaxies are likely significantly higher than previously thought. In particular, in their recent *HST* study of novae in M87 Michael Shara and collaborators (Shara et al. 2016) have suggested that previous ground-based, relatively low-cadence surveys, may have systematically underestimated the LSNRs of most galaxies. It is quite straightforward to see how this could happen. Due to the practical difficulty in securing long blocks of observing time, most ground-based nova surveys have been done with a low cadence. In order to compute a nova rate, it was necessary to make rather significant corrections in extrapolating the observed number of novae into an annual rate. To accomplish this correction, Monte Carlo simulations using template nova light curves are typically employed to estimate, for a given annual nova rate, the fraction of novae visible in a survey with a given temporal sampling. Thus, the properties of the template light curves are a critical parameter in estimating the nova rates. In the Hα imaging studies of Shafter and collaborators, template light curves have been drawn from well-sampled nova light curves from M31. These templates were usually selected because they were the most complete Hα light curves available, but there has always been a worry that they may not accurately reflect the intrinsic nova luminosity function of novae in a given galaxy, and perhaps not even in M31 itself.

In addition to the uncertainties that result through limited temporal coverage, the computation of the LSNR$_K$ can also introduce uncertainty. In the vast majority of cases, nova imaging surveys do not cover the full spatial extent of the Galaxy being studied. As a result, a determination of the overall nova rate requires an extrapolation of the nova rate estimates within the *surveyed* region to the entire galaxy. In many cases, the LSNRs were determined after estimating the global nova rate and normalizing by an estimate of the integrated infrared (usually K-band) luminosity of the galaxy. An alternative, and more robust option for computing the $LSNR_K$ is simply to normalize the nova rate in the surveyed region by the K-band luminosity *within the surveyed region.* The latter method has the advantage that it avoids extrapolation errors; however, the derived $LSNR_K$ may not accurately reflect that of the Galaxy as a whole. The two methods will yield similar results if the specific nova rate does not vary spatially within a galaxy and if the integrated galaxy luminosity has been reliably measured.

An example where the computation of the $LSNR_K$ can be quite complicated concerns the Virgo cluster elliptical galaxy, M87. The field-of-view of the recent archival *HST* studies of M87 (Shara et al. 2016; Shafter et al. 2017) covered an area of only 202″ × 202″, which represents less than 50% of the Galaxy's integrated light. As a result, a determination of the global M87 nova rate, and a $LSNR_K$ based on it, requires an extrapolation to the observed rate, and the assumption that the $LSNR_K$ does not vary within M87. Although, Shara et al. (2016) determined that the nova density appeared to follow the background light within the area covered by the *HST* observations, a pioneering study published a decade earlier by Juan Madrid and collaborators (Madrid et al. 2007) had suggested that the nova rate in the central region of M87 may be considerably higher than that of the Galaxy overall. In particular, Madrid et al. found a total of 13 transient events consistent with nova eruptions, corresponding to a nova rate of ~64 yr^{-1} within a 24.7″ × 24.7″ region centered on the nucleus of M87. Given that this region covers only ~6% of the Galaxy's light, a direct extrapolation under the assumption that the nova density follows the background light led to a global rate of well over 1000 novae per year—a value that is ruled out by ground-based observations that cover the full spatial extent of M87 (Shafter et al. 2000). Clearly, if the transients discovered by Madrid et al. are in fact *bona fide* novae, the $LSNR_K$ must vary spatially within M87. Another issue with extrapolation of the light of M87 is distinguishing the extended halo of the Galaxy from the intracluster medium. The overarching lesson here is that, in the case of a galaxy like M87, an extrapolation of the $LSNR_K$ in the central region of the Galaxy may significantly overestimate that of the Galaxy as a whole.

At this point, no definitive conclusions can be drawn concerning whether the $LSNR_K$ varies with the Hubble type of the host galaxy. All we can say is that, within the uncertainties of the measurements for the individual galaxies, there is at present no compelling evidence that the $LSNR_K$ varies systematically across the Hubble sequence. This uncertainty has been with us since the first studies were undertaken in the 1990s, and it is unlikely to be lifted until deep, high-cadence surveys have been applied to a wide range of galaxies. Only then will we be able to explore the full rage of luminosities and fade rates in a given nova population that will be necessary to accurately measure extragalactic nova rates.

8.6 Final Thoughts

The last 100 years has seen a dramatic advance in our understanding of Classical Nova systems. We are rapidly approaching the centennial of the Great Debate on the nature of the spiral nebulae. At the time of the Debate, the distinction between novae and supernovae was not yet appreciated, the physical nature of both was completely unknown, and the existence of external galaxies was still very much still a matter of conjecture. Within the next 50 years these issues would be resolved. By 1930, largely through the work of Edwin Hubble, the external galaxies would be recognized, and the size of the Universe would become unimaginably vast. Within the next decade, the distinction of novae and supernovae would be established, with the fundamental nature of supernovae revealed. The next 30 years would reveal that novae are the result of a thermonuclear runaway on the surface of an accreting white dwarf in a close binary system.

Over the last 50 years, despite tremendous advances in technology, progress in understanding the nova phenomena has been less dramatic, with many questions remaining unanswered. Are there two populations of nova progenitors, and if so, how do they affect nova properties across differing stellar populations? The MMRD is currently on "life-support" given the observations of Kasliwal et al. (2011) and Schaefer (2018), but recent work by Selvelli & Gilmozzi (2019) suggest that it may be too soon to write its obituary. Are there additional parameters yet to be elucidated that may reduce the scatter in this beleaguered relation? How common are RNe, and in particular, the rapidly-recurring systems like M31N 2008-12a? The answer to the latter question has clear implications for the viability of novae as a significant progenitor channel for SNe Ia. The key to resolving these and similar questions lies in exploiting the wealth of data that will soon come from deep and high-cadence surveys, both planned (e.g., the Large Synoptic Survey Telescope; Ivezić et al. 2019) and currently underway (e.g., the Zwicky Transient Facility; Bellm et al. 2018).

References

Barsukova, E., Fabrika, S., Hornoch, K., et al. 2011, ATel #3725
Bellm, E. C., Kulkarni, S. R., Graham, M. J., et al. 2018, PASP, 131, 018002
Bode, M. F., Darnley, M. J., Beardmore, A. P., et al. 2016, ApJ, 818, 145
Buscombe, W., & de Vaucouleurs, G. 1955, Obs, 75, 170
Ciardullo, R., Ford, H. C., Neill, J. D., et al. 1987, ApJ, 318, 520
Cohen, J. G. 1985, ApJ, 292, 90
Darnley, M. J., Williams, S. C., Bode, M. F., et al. 2014, A&A, 563, 9
Della Valle, M., Rosino, L., Bianchini, A., & Livio, M. 1994, A&A, 287, 403
Della Valle, M., & Livio, M. 1995, ApJ, 452, 704
Downes, R. A., & Duerbeck, H. W. 2000, AJ, 120, 2007
Freedman, W. L., Madore, B. F., Gibson, B. K., et al. 2001, ApJ, 553, 47
Henze, M., Ness, J.-U., Darnley, M. J., et al. 2014, A&A, 563, 8
Hillman, Y., Prialnik, D., Kovetz, A., & Shara, M. M. 2016, ApJ, 819, 168
Hornoch, K., & Shafter, A. W. 2015, ATel #7116
Ivezić, Z., Kahn, S. M., Tyson, J. A., et al. 2019, ApJ, 873, 111

Kasliwal, M. M., Cenko, S. B., Kulkarni, S. R., et al. 2011, ApJ, 735, 94
Kato, M., Saio, H., Hachisu, I., & Nomoto, K. 2014, ApJ, 793, 136
Leavitt, H. S., & Pickering, E. C. 1912, HarCi, 173, 1
Lindegren, L., Hernández, J., Bombrun, A., et al. 2018, A&A, 616, 2
Madrid, J. P., Sparks, W. B., Ferguson, H. C., et al. 2007, ApJ, 654, 41
McLaughlin, D. B. 1945, PASP, 57, 69
McLaughlin, D. B. 1960, in Stellar Atmospheres, ed. J. L. Greenstein (Chicago, IL: University of Chicago Press), 585
Mróz, P., Udalski, A., Poleski, R., et al. 2016, ApJS, 222, 9
Mróz, P., & Udalski, A. 2018, ATel #11384
Özdönmez, A., Güver, T., Cabrera-Lavers, A., & Ak, T. 2016, MNRAS, 461, 1177
Özdönmez, A., Ege, E., Güver, T., & Ak, T. 2018, MNRAS, 476, 4162
Pagnotta, A., & Schaefer, B. E. 2014, ApJ, 788, 164
Paxton, B., Bildsten, L., Dotter, A., et al. 2011, ApJS, 192, 3
Paxton, B., Cantiello, M., Arras, P., et al. 2013, ApJS, 208, 4
Pietsch, W. 2010, AN, 331, 187
Rosino, L. 1973, A&AS, 9, 347
Schaefer, B. E. 2010, ApJS, 187, 275
Schaefer, B. 2018, MNRAS, 481, 3033
Schmidt, T. 1957, VeGoe, 7, 1
Selvelli, P., & Gilmozzi, R. 2013, A&A, 560, A49
Selvelli, P., & Gilmozzi, R. 2019, A&A, 622, A168
Shafter, A. W. 1997, ApJ, 487, 226
Shafter, A. W., Ciardullo, R., & Pritchet, C. J. 2000, ApJ, 530, 193
Shafter, A. W., & Irby, B. K. 2001, ApJ, 563, 749
Shafter, A. W., Ciardullo, R., Bode, M. F., et al. 2008, ATel #1834
Shafter, A. W., Rau, A., Quimby, R. M., et al. 2009, ApJ, 690, 1148
Shafter, A. W., Darnley, M. J., Hornoch, K., et al. 2011a, ApJ, 734, 12
Shafter, A. W., Ciardullo, R., & Darnley, M. J. 2011b, ATel #3727
Shafter, A. W. 2011, ATel #3728
Shafter, A. W., Hornoch, K., Ciardullo, R., et al. 2012, ATel #4503
Shafter, A. W. 2013, AJ, 145, 117
Shafter, A. W., Henze, M., Rector, T. A., et al. 2015, ApJS, 216, 34
Shafter, A. W. 2017, ApJ, 834, 196
Shafter, A. W., Henze, M., Darnley, M. J., et al. 2017, RNAAS, 1, 44
Shara, M. M., Doyle, T. F., Lauer, T. R., et al. 2016, ApJS, 227, 1
Shara, M. M., Doyle, T. F., Pagnotta, A., et al. 2018, MNRAS, 474, 1746
Shida, R. Y., & Liller, W. 2004, ASPC, 310, 184
Sin, P., Henze, M., & Sala, G. 2017, ATel #10001
Socia, Q., Henze, M., Shafter, A. W., & Horst, J. C. 2018, RNAAS, 2, 190
Soraisam, M. D., & Gilfanov, M. 2015, A&A, 583, 140
Soraisam, M. D., Gilfanov, M., Kupfer, T., et al. 2017, A&A, 599, 48
Starrfield, S., Timmes, F. X., Iliadis, C., et al. 2012, BaltA, 21, 76
Strope, R. J., Schaefer, B. E., & Henden, A. A. 2010, AJ, 140, 34
Tang, S., Cao, Y., & Kasliwal, M. M. 2013, ATel #5607
Tang, S., Bildsten, L., Wolf, W. M., et al. 2014, ApJ, 786, 61

van den Bergh, S. 2000, in Cambridge Astrophysics Series 35, The Galaxies of the Local Group (Cambridge: Cambridge University Press)
Williams, S. C., Darnley, M. J., Bode, M. F., & Shafter, A. W. 2013, ATel #5611
Wolf, W. M., Bildsten, L., Brooks, J., & Paxton, B. 2013, ApJ, 777, 136
Yaron, O., Prialnik, D., Shara, M. M., & Kovetz, A. 2005, ApJ, 623, 398
Yungelson, L., Livio, M., & Tutukov, A. 1997, ApJ, 481, 127

Lightning Source UK Ltd.
Milton Keynes UK
UKHW030626011119
352580UK00014B/310/P

9 780750 318464